SEAWATER: ITS COMPOSITION, PROPERTIES AND BEHAVIOUR

THE OCEANOGRAPHY COURSE TEAM

Authors
Evelyn Brown (*Waves, Tides, etc.; Ocean Chemistry*)
Angela Colling (*Ocean Circulation; Case Studies*)
Dave Park (*Waves, Tides, etc.*)
John Phillips (*Case Studies*)
Dave Rothery (*Ocean Basins*)
John Wright (*Ocean Basins; Seawater; Ocean Chemistry; Case Studies*)

Designer
Jane Sheppard

Graphic Artist
Sue Dobson

Cartographer
Ray Munns

Editor
Gerry Bearman

This Volume forms part of an Open University course. For general availability of all the Volumes in the Oceanography Series, please contact your regular supplier, or in case of difficulty the appropriate Butterworth-Heinemann office.

Further information on Open University courses may be obtained from: The Admissions Office, The Open University, P.O. Box 48, Walton Hall, Milton Keynes MK7 6AA, UK.

Cover illustration: Satellite photograph showing distribution of phytoplankton pigments in the North Atlantic off the US coast in the region of the Gulf Stream and the Labrador Current. (*NASA, and O. Brown and R. Evans, University of Miami.*)

SEAWATER: ITS COMPOSITION, PROPERTIES AND BEHAVIOUR

PREPARED BY AN OPEN UNIVERSITY COURSE TEAM

in association with

The Open University

THE OPEN UNIVERSITY, WALTON HALL,
MILTON KEYNES, MK7 6AA, ENGLAND

Butterworth-Heinemann
Linacre House, Jordan Hill, Oxford OX2 8DP
A division of Reed Educational and Professional Publishing Ltd

⟨ A member of the Reed Elsevier plc group

OXFORD   BOSTON   JOHANNESBURG
MELBOURNE   NEW DELHI   SINGAPORE

British Library Cataloguing in Publication Data
A catalogue record for this book is available from the British Library

ISBN 0 7506 3715 3

Library of Congress Cataloguing in Publication Data
A catalogue record for this book is available from the Library of Congress

Jointly published by the Open University, Walton Hall, Milton Keynes
MK7 6AA and Butterworth-Heinemann

Edited, typeset, illustrated and designed by The Open University
Printed in Singapore by Kyodo under the supervision of MRM
Graphics Ltd., UK

s330v2i2.4

CONTENTS

ABOUT THIS VOLUME

This is one of a Series of Volumes on Oceanography. It is designed so that it can be read on its own, like any other textbook, or studied as part of S330 *Oceanography*, a third level course for Open University students. The science of oceanography as a whole is multidisciplinary. However, different aspects fall naturally within the scope of one or other of the major 'traditional' disciplines. Thus, you will get the most out of this Volume if you have some previous experience of studying chemistry and a certain amount of physics. Other Volumes in this Series lie variously within the fields of geology, biology, physics or chemistry.

Chapter 1 summarizes the special properties of water and the role of the oceans in the hydrological cycle. Chapters 2 to 4 discuss the distribution of temperature and salinity in the oceans and their combined influence on density, stability and vertical water movements. Chapter 5 describes the behaviour of light and sound in seawater and provides examples of the application of acoustics to oceanography. Chapter 6 examines the composition and behaviour of the dissolved constituents of seawater, covering both minor and trace constituents and the major ions, as well as dissolved gases and biologically important nutrients. It deals also with such topics as residence times, speciation and carbonate equilibria. Finally, Chapter 7 provides a short review of ideas about the history of seawater, the involvement of the oceans in global cycles and their relationship to climatic change.

You will find questions designed to help you to develop arguments and/or test your own understanding as you read, with answers provided at the back of this Volume. Important technical terms are printed in **bold** type where they are first introduced or defined.

ABOUT THIS SERIES

The Volumes in this Series are all presented in the same style and format, and together provide a comprehensive introduction to marine science. *Ocean Basins* deals with structure and formation of oceanic crust, hydrothermal circulation, and factors affecting sea-level. *Seawater* considers the seawater solution and leads naturally into *Ocean Circulation*, the 'core' of the Series, providing a largely non-mathematical treatment of ocean–atmosphere interaction and the dynamics of wind-driven surface current systems, and of density-driven circulation in the deep oceans. *Waves, Tides and Shallow-Water Processes* introduces processes controlling water movement and sediment transport in nearshore environments (beaches, estuaries, deltas, shelves). *Ocean Chemistry and Deep-Sea Sediments* is concerned with biogeochemical cycling of elements within the seawater solution and with water–sediment interaction on the ocean floor. *Case Studies in Oceanography and Marine Affairs* examines the effect of human intervention in the marine environment and introduces the essentials of Law of the Sea. The two interdisciplinary case studies respectively review marine affairs in the Arctic from an historical standpoint, and outline the causes and effects of the tropical climatic phenomenon known as El Niño.

Biological Oceanography: An Introduction (by C. M. Lalli and T. R. Parsons) is a companion Volume to the Series, and is also in the same style and format. It describes and explains interactions between marine plants and animals in relation to the physical/chemical properties and dynamic behaviour of the seawater in which they live.

CHAPTER 1   WATER, AIR AND ICE

The oceans and atmosphere originated mostly from inside the Earth. Water vapour and other gases have been progressively released from the Earth's interior by a process of de-gassing that has been going on since the Earth formed about 4.6 billion years ago. The rate of de-gassing has decreased through time, because the radioactive elements responsible for much of the Earth's internal heat have been decaying exponentially, and there is much less of these elements now than there was when the Earth formed. In short, the interior of the early Earth was hotter than it is now; convection in the Earth's mantle was more vigorous; and de-gassing was more rapid. It seems likely that most of the water and atmospheric gases originally inside the Earth had been de-gassed by about 2.5 billion years ago, and that de-gassing has continued ever since but at a progressively decreasing rate. Small amounts of water and atmospheric gases continue to be expelled from the Earth's interior even today.

The oceans and atmosphere together provide our fluid environment. The nature of that environment is controlled to a very large extent by the special properties of a substance we take virtually for granted: water.

1.1   THE SPECIAL PROPERTIES OF WATER

'From a drop of water, a logician could infer the possibility of an Atlantic or a Niagara, without having seen or heard of one or the other.'

Sherlock Holmes, in *A Study in Scarlet*, by Sir Arthur Conan Doyle.

It is easy enough, perhaps, to infer the *existence* of oceans from a drop of water, less easy to deduce that they have waves, tides and currents, still less easy to predict patterns of water movement and water chemistry, and the nature of marine life forms. Nonetheless, a knowledge of the properties of water does enable us to understand at least some of the major characteristics of the oceanic environment.

QUESTION 1.1   Most people know that the oceans are salty, cold, dark and teem with noisy life, and that they are never still. Explain these characteristics of the oceans by selecting items from the following list of properties and attributes of water.

Water is a highly mobile liquid

Water is a good solvent

Water is a poor conductor of heat

Water has a high specific heat

Water has high latent heats of fusion and of evaporation

Pure water freezes at $0\,°C$

Pure water boils at $100\,°C$

The maximum density of freshwater is at $4\,°C$; for seawater, it is at its freezing point ($-1.9\,°C$)

Ice is less dense than water

Light can only travel a maximum of a few hundred metres through water

Sound can travel thousands of kilometres through water

Water is essential to life

The molecular mass of water is 18. Comparison with other hydrogen compounds of comparable molecular mass suggests that water should freeze at about $-100\,°C$ and boil at about $-80\,°C$, instead of at $0\,°C$ and $100\,°C$ respectively (e.g. methane, with a molecular mass of 16, freezes at $-183\,°C$ and boils at $-162\,°C$). The density of most solids is greater than that of their corresponding liquids, and the density of liquids typically decreases progressively when heated from the melting point – but ice is less dense than water, and the maximum density of pure water is reached at $4\,°C$. Tables 1.1 and 1.2 (overleaf) contain much the same information as that summarized for Question 1.1, but in a more detailed and quantitative form.

The reasons for these anomalous properties of water lie in its molecular structure. A water molecule consists of an oxygen atom bonded to two hydrogen atoms. The angle between the interatomic bonds is $105°$. The difference in electrical properties between oxygen and hydrogen atoms results in the hydrogen side carrying a small positive charge, while the oxygen atom carries a small negative charge (Figure 1.1). Because of this polar structure, water molecules have an attraction for one another and tend to arrange themselves into partially ordered groups, linked by weak intermolecular bonds called hydrogen bonds.

As the temperature of pure liquid water is raised above $0\,°C$, the energy of the molecules increases, counteracting the tendency to form partially ordered groups. Individual molecules can then fit together more closely, occupying less space and increasing the density of the water. However, further raising the temperature imparts yet more energy to the molecules, and the average distance between them increases, which results in decreased density. At temperatures between $0\,°C$ and $4\,°C$, the 'ordering effect' predominates, whereas at higher temperatures thermal expansion is more important. The combination of the two effects means that the density of pure water is greatest at $4\,°C$ (Table 1.2).

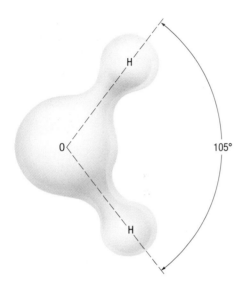

Figure 1.1 Schematic view of the water molecule. It is electrically polarized. The oxygen side carries a small negative charge; the hydrogen side carries a small positive charge.

1.1.1 THE EFFECT OF DISSOLVED SALTS

Any substance dissolved in a liquid has the effect of increasing the density of that liquid. The greater the amount dissolved, the greater the effect. Water is no exception. The density of freshwater is close to $1.00 \times 10^3\,kg\,m^{-3}$ (cf. Table 1.2), while the average density of seawater is about $1.03 \times 10^3\,kg\,m^{-3}$.

Another important effect of dissolved substances is to depress the freezing point of liquids. For example, the addition of common salt (sodium chloride, NaCl) lowers the freezing point of water – which is why salt is spread on frozen roads. It also lowers the temperature at which water reaches its maximum density. That is because dissolved salts inhibit the tendency of water molecules to form ordered groups, so that density is controlled only by the thermal expansion effect. Figure 1.2 shows that the freezing point and the temperature of maximum density are the same when the concentration of dissolved salts in water (the salinity) reaches about $25\,g\,kg^{-1}$. The oceans have higher salinity than this, about $35\,g\,kg^{-1}$ on average (of which about $30\,g\,kg^{-1}$ are contributed by dissolved sodium ions (Na^+, ~ 11 g) and chloride ions (Cl^-, ~ 19 g)). Therefore, the density of seawater increases with falling temperature right down to the freezing point. This is a crucial distinction between freshwater and seawater and it has a profound effect on the formation of sea-ice and on oceanic circulation processes, as you will see in later Chapters.

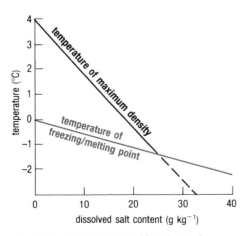

Figure 1.2 Temperatures of freezing and melting point and maximum density of liquid water as functions of the concentration of dissolved salts.

Table 1.1 Anomalous physical properties of liquid water. (You are *not* expected to remember details of this Table.)

Property	Comparison with other substances	Importance in physical/biological environment
Specific heat ($= 4.18 \times 10^3$ J kg^{-1} °C^{-1})	Highest of all solids and liquids except liquid NH$_3$	Prevents extreme ranges in temperature; heat transfer by water movements is very large; tends to maintain uniform body temperatures
Latent heat of fusion ($= 3.33 \times 10^5$ J kg^{-1})	Highest except NH$_3$	Absorption or release of latent heat results in large thermostatic effect at freezing point
Latent heat of evaporation ($= 2.25 \times 10^6$ J kg^{-1})	Highest of all substances	Absorption or release of latent heat results in large thermostatic effect at boiling point; large latent heat of evaporation is extremely important in heat and water transfer within the atmosphere
Thermal expansion	Temperature of maximum density decreases with increasing salinity; for pure water it is at 4 °C	Freshwater and dilute seawater have maximum density at temperatures above the freezing point; the maximum density of normal seawater is at the freezing point
Surface tension ($= 7.2 \times 10^9$ N m^{-1}) *	Highest of all liquids	Important in cell physiology; controls certain surface phenomena and the formation and behaviour of droplets
Dissolving power	In general, dissolves more substances and in greater quantities than any other liquid	Obvious implications in both physical and biological phenomena
Dielectric constant † ($= 87$ at 0 °C, 80 at 20 °C)	Pure water has the highest of all liquids except H$_2$O$_2$ and HCN	Important in the behaviour of inorganic dissolved substances because of the resulting high dissociation
Electrolytic dissociation	Very small	A neutral substance, yet contains both H$^+$ and OH$^-$ ions
Transparency	Relatively great	Absorption of radiant energy is large in infrared and ultraviolet; in the visible portion of the energy spectrum there is relatively little selective absorption, hence pure water is 'colourless' in small amounts; characteristic absorption important in physical and biological phenomena
Conduction of heat	Highest of all liquids	Important on a small scale, as in living cells, but molecular processes outweighed by turbulent diffusion
Molecular viscosity ($= 10^{-3}$ N s m^{-2}) *	Less than most other liquids at comparable temperature	Flows readily to equalize pressure differences

* N = newton = unit of force in kg m s^{-2}.

† Measure of the ability to keep oppositely charged ions in solution apart from one another.

Notes to Table 1.1

1   Latent heat is the amount of heat required to melt unit mass of a substance at the melting point, or to evaporate unit mass of a substance without change of temperature.

2   Specific heat is the amount of heat required to raise the temperature of unit mass of a substance by one degree.

3   Surface tension is a measure of the 'strength' of the liquid surface and hence of the 'durability' of drops and bubbles (see Section 2.2.1).

4   Viscosity is a measure of resistance to distortion (i.e. flow) of a fluid. The greater the viscosity, the less readily will the fluid flow (e.g. motor oil is more viscous than water).

Table 1.2 Density of pure water at different temperatures.

Temperature (°C)	State	Density (kg m^{-3})
−2	Solid	917.2
0	Solid	917.0
0	Liquid	999.8
4	Liquid	1 000.0
10	Liquid	999.7
25	Liquid	997.1

QUESTION 1.2 On Figure 1.2 (p. 5), do the words 'maximum density' refer to a single density value, or does the maximum density itself increase or decrease along the line, with falling temperature and increasing dissolved salt content?

1.2 THE HYDROLOGICAL CYCLE

Figure 1.3   The hydrological cycle, showing the Earth's water inventory, annual movements of water through the cycle (black numbers), and amounts of water stored in different parts of the cycle (blue numbers). Note the overwhelming dominance of ocean water in the inventory. All quantities shown are × 10^{15} kg. Note also that 10^{15} kg water ≈ 10^3 km³.

The oceans dominate the **hydrological cycle** (Figure 1.3), for they contain 97% of the global water inventory. Large changes in terrestrial parts of the water inventory (Table 1.3) would be necessary to have any significant effect on the amount of water in the oceans. For example, it is estimated that during the glacial maxima of the past two million years, some 50 000 × 10^{15} kg of water were added to the world's glaciers and ice-caps, increasing their volume to about two-and-a-half times what it is today.

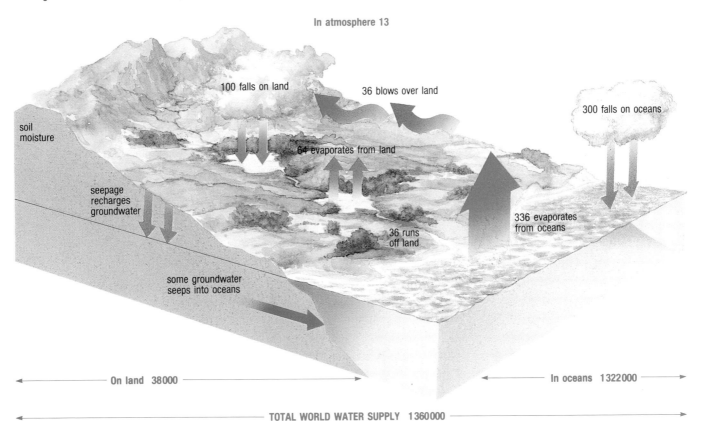

This lowered sea-level world-wide by over 100 m – enough to turn most shallow continental-shelf seas into dry land – but it only reduced the total volume of water in the oceans by about 3.5%.

The concept of **residence time** can be defined by reference to Figure 1.3. It is the average length of time that a water molecule resides – or is stored – in any particular stage of the hydrological cycle. It is calculated by dividing the amount of water in that part of the cycle by the amount that enters (and leaves) it in unit time. (You will encounter residence time in other contexts later.)

Table 1.3   Water on land (× 10^{15} kg).

Rivers and streams	1
Freshwater lakes	125
Salt lakes and inland seas	104
Total surface water	230
Glaciers and ice-caps	29 300
Soil moisture and seepage	70
Groundwater	8 400
Total on land	38 000

QUESTION 1.3 (a)   Look at Figure 1.3. What is the annual rate of evaporation from the ocean? Is it balanced by precipitation plus run-off from land?

(b)   What is the residence time of water in the oceans?

(c)   Approximately what quantity of water moves through the atmosphere annually?

1.2.1   WATER IN THE ATMOSPHERE

The most obvious manifestations of water in the atmosphere are clouds and fog. Both consist of water droplets or ice crystals that have condensed round (or nucleated on) small particles in the air. Water in the atmosphere is mostly in the gaseous state, i.e. as water vapour. Air is saturated with water vapour when there is equilibrium between evaporation and condensation. The higher the temperature, the greater the amount of energy available for evaporation, so warm air can hold more moisture at saturation (i.e. it has higher humidity) than cold air.

There are two ways in which unsaturated air can be cooled so that it becomes saturated and condensation begins:

1   Cooling occurs when air rises and expands adiabatically as atmospheric pressure decreases with height. **Adiabatic changes** of temperature are those that occur independently of any transfer of heat to or from the surroundings (see also Section 4.2.1). Thus, rising air expands and loses internal energy, so that its temperature may fall sufficiently for the water vapour it contains to condense as water droplets and form cloud or fog.

2   Cooling also occurs when air comes into contact with a cold surface (which is why you get condensation on windows in winter, for example).

Fogs develop when a sufficiently thick layer of moist air is cooled to condensation point, forming in effect clouds at ground (or water) level. Two main types of fog are recognized.

Radiation fog forms when the ground surface is cooled by radiant heat loss at night into a clear sky. If the air in contact with the ground is close to saturation and its temperature falls sufficiently, then fog may form. Radiation fogs do not develop over lakes or the sea, because water has a high specific heat (Table 1.1), so water surfaces cool less rapidly than ground surfaces. However, radiation fog often drifts from land over rivers, estuaries and coastal waters.

Advection fog forms when warm humid air moves (is advected) over cold ground or water and is cooled. Such fogs commonly develop over the Grand Banks off Newfoundland for example, where air formerly above the warm Gulf Stream is advected over the cold Labrador Current (see Figure 2.11).

1.2.2   ICE IN THE OCEANS

Polar ice-caps are a significant feature of the present-day Earth. A layer of ice with its covering of snow reflects back more incoming solar radiation than areas of land or open water (see Section 2.1). Only a small amount of solar energy can penetrate to surface water or land under the ice, so that once it is established, ice tends to be self-perpetuating.

Sea-ice is formed by the freezing of seawater itself. Various stages and ages of sea-ice formation are illustrated in Figure 1.4. When seawater first begins to freeze, relatively pure ice is formed, so that the salt content of the surrounding seawater is increased, which both increases its density and depresses its freezing point further (cf. Figure 1.2). Most of the salt in sea-ice is in the form of concentrated brine droplets trapped within the ice as it forms. Brine trapped in sea-ice is much more saline than the ice itself. Its freezing point is greatly depressed relative to that of the ice (cf. Figure 1.2); and so the brine droplets remain liquid at temperatures well below those of ice formation.

QUESTION 1.4   (a)   Give two reasons why these brine droplets will be denser than the surrounding ice.

(b)   Old sea-ice is less saline than young sea-ice formed under comparable conditions, i.e. it contains fewer brine droplets. Can you explain how that happens?

Round the South Pole the centre of ice accumulation is the ice-buried continent of Antarctica, which is surrounded by a shelf of floating land ice, the outer part of the continental ice-cap. In contrast, the North Pole is surrounded by the basin of the Arctic Ocean, which is largely covered by floating sea-ice. In both regions, floating ice-fields are called pack-ice (Figure 1.4).

(a)

(b)

(c)

(d)

Figure 1.4   Forms of sea-ice.

(a)   Grease ice or frazil ice, formed as ice crystals grow and coalesce to give the surface an oily appearance.

(b)   Pancake ice, formed as grease ice thickens and breaks up into pieces, usually about 0.5–3 m across. The roughly circular shapes and raised rims result from continued collision with one another.

(c)   Open pack-ice, formed of ice-floes with many channels (leads) between them.

(d)   Close pack-ice, formed of ice-floes mostly in contact, with few leads.

(e)   Brash ice, fragments not more than about 2 m across, the 'wreckage' of other forms of ice, here seen stranded at a shoreline.

(e)

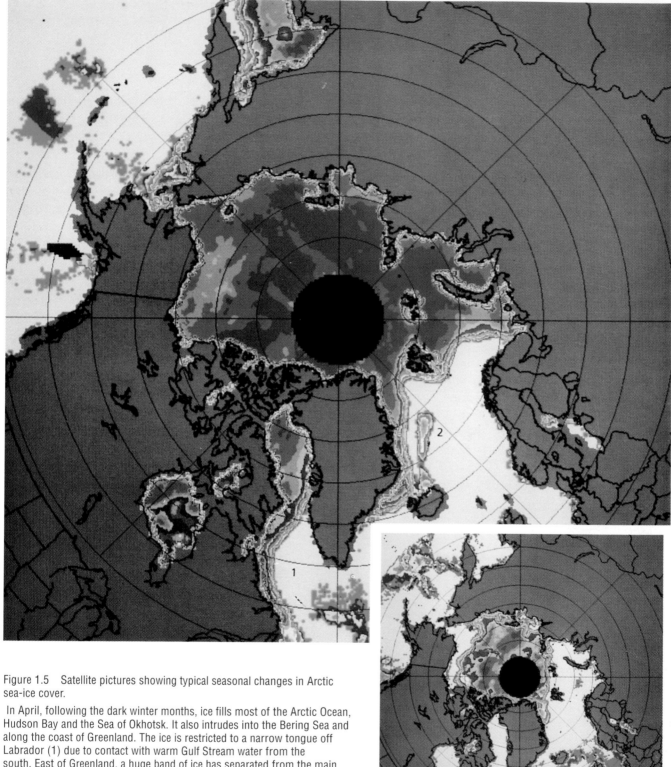

Figure 1.5   Satellite pictures showing typical seasonal changes in Arctic sea-ice cover.

In April, following the dark winter months, ice fills most of the Arctic Ocean, Hudson Bay and the Sea of Okhotsk. It also intrudes into the Bering Sea and along the coast of Greenland. The ice is restricted to a narrow tongue off Labrador (1) due to contact with warm Gulf Stream water from the south. East of Greenland, a huge band of ice has separated from the main ice pack (2).

(*Inset*): In September, following warmer months of nearly continual daylight, the pack ice has melted and receded to the confines of the Arctic Ocean.

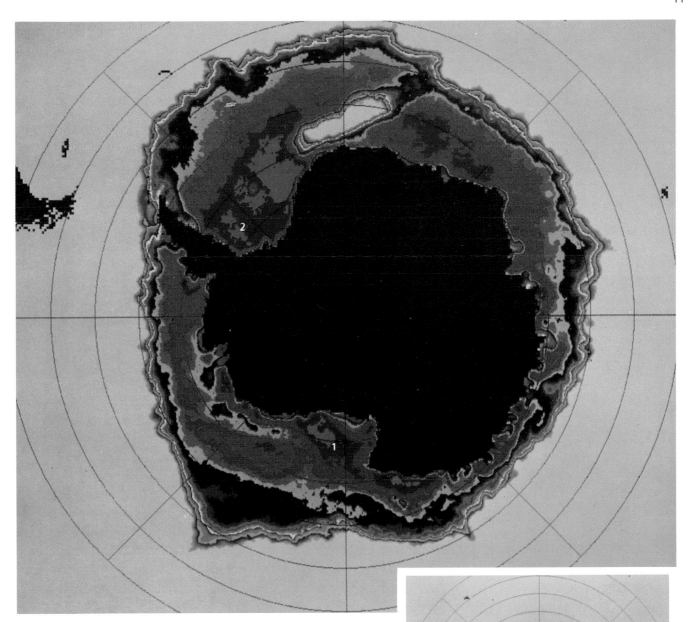

Figure 1.6   Satellite pictures showing typical seasonal changes in Antarctic sea-ice cover.

In August (the southern winter), the ice cover surrounds Antarctica, extending more than 1 000 km from the continent into the Ross (1) and Weddell (2) Seas. The large ice-free enclosure seen in the eastern part of the Weddell Sea is of particular interest. It does not form every year, but when it does, it is found in approximately the same position. Why it forms is not fully understood, but it must be due to warm water rising from below or it would not persist through the winter.

(*Inset*): By February, sustained heating during continuous summer daylight has melted more than 80% of the winter sea-ice cover, and ice extent is at its summer minimum.

Seasonal and inter-annual changes in ice cover at high latitudes are monitored by satellite (Figures 1.5 and 1.6). A knowledge of the extent of ice cover and the way it changes with time is crucial to understanding and predicting weather patterns and such data are important for assessing the surface radiation balance of the Earth as a whole.

Icebergs form in different ways in the two hemispheres. In the Arctic, they are mostly brought to the sea by valley glaciers from land masses such as Greenland and Spitsbergen, and they are irregular in shape. Thick ice-sheets fringe parts of the Arctic Ocean and large areas occasionally break off to form ice islands, upon which scientific bases have been established to make observation platforms that drift around the Arctic Ocean. In the Southern Ocean, tabular icebergs are more typical, and are formed by 'calving' from the ice-shelves of the Ross and Weddell Seas (Figure 1.7).

Arctic icebergs contain more soil and debris eroded from the land by the glaciers and are usually denser and darker in colour than those of the Antarctic. They are also smaller, being rarely more than 1 km long, though

(a)

Figure 1.7 (a) Seaward-facing cliffs of part of the Antarctic ice-shelf. The flow of ice from the land causes the ice-front to advance, but this is balanced by the 'calving' of (b) tabular icebergs from the cliffs.

(b)

they can be as high as 60 m above the sea-surface. Antarctic icebergs may have surface areas of many square kilometres, but they are rarely more than 35 m high. As icebergs melt, they dilute the surface seawater with freshwater, and the salinity of surface seawater in high latitudes is appreciably less than in ice-free latitudes: about 30–33 g kg⁻¹, as against 35 g kg⁻¹.

The next Chapter is concerned with the distribution of temperature in the oceans and some of the reasons for that distribution.

1.3 SUMMARY OF CHAPTER 1

1 The special properties of water – in particular, its anomalously high melting and boiling points, specific and latent heats, powerful solvent properties, and maximum density at 4 °C – result from the polar structure of the water molecule. Dissolved salts increase the density of water and depress both the temperature of maximum density and the freezing point.

2 The oceans contain 97% of the water that circulates in the hydrological cycle. The residence time of water in the oceans is measured in thousands of years; in the atmosphere, it is measured in days.

3 Air is saturated with water vapour when evaporation is balanced by condensation. Clouds and fog are condensed water vapour. Fog may form when air is cooled to its condensation temperature, either by radiation from the land, or by advection of warm humid air over a cool land or water surface.

4 Sea-ice is less saline than the seawater from which it freezes, so its formation increases the salt content of the remaining seawater, thus further depressing its freezing point and increasing its density. Icebergs in the Northern Hemisphere are formed when valley glaciers on lands surrounding the Arctic Ocean reach the sea; those in the Southern Hemisphere break off from the thick ice-shelf that surrounds the Antarctic continent.

Now try the following questions to consolidate your understanding of this Chapter.

QUESTION 1.5 (a) In what ways are the thermal properties of water probably the single most important factor in preventing extremes of temperature from being reached at the Earth's surface?

(b) Most liquids reach a maximum density at freezing point, but pure water is an exception. At what temperature does pure water reach maximum density? Does this temperature apply to seawater?

QUESTION 1.6 What is the approximate average residence time of water on land, and why is this average value likely to conceal considerable variations?

QUESTION 1.7 (a) What is the main difference in origin between the ice-sheets that cover the Arctic and Antarctic polar regions?

(b) A sample of water contains 20 g kg⁻¹ dissolved salts. At what temperature will it (i) attain its maximum density, (ii) freeze?

(c) Ice melts and mixes with seawater of salinity 35 g kg⁻¹. Will this have the effect of raising or lowering the freezing point of the seawater? Would this in turn tend to facilitate the formation of more sea-ice when temperatures fell once more?

CHAPTER 2   TEMPERATURE IN THE OCEANS

Two of the most important properties of seawater are temperature and salinity (concentration of dissolved salts), for together they control its density, which is the major factor governing the vertical movement of ocean waters.

In the oceans, the density of seawater normally increases with depth. If the density of surface water exceeds that of the underlying water, the situation is gravitationally unstable and the surface water sinks. In polar regions, the density of surface waters can be increased in two ways: first, by direct cooling, either where ice is in contact with the water or where cold winds blow off the ice; secondly, by the formation of sea-ice, which extracts water and leaves behind seawater of higher salinity and increased density (Section 1.2.2). The cold dense currents of the deep circulation (see Section 4.1) originate by sinking of dense water in polar regions. In lower latitudes, dense saline water is produced by excess evaporation, which may be aided by strong winds such as those that occur during the winter in parts of the Mediterranean.

2.1 SOLAR RADIATION

The Sun's radiation is dominated by ultraviolet, visible and near infrared wavelengths (see Section 2.3). On average, only about 70% of the solar radiation that reaches the Earth penetrates the atmosphere. About 30% (on average) is *reflected* back into space from clouds and dust particles. Of the remaining 70%, on average:

about 17% is *absorbed* in the atmosphere;

about 23% reaches the surface as *diffuse* daylight;

about 30% reaches the surface as *direct* sunlight.

Much of the ultraviolet radiation is absorbed in the ozone layer. (A cloudless sky appears blue because of scattering of shorter wavelengths by molecules of atmospheric gases.)

The radiation that actually reaches the Earth's surface – the **insolation** – is not all absorbed. The percentage of the insolation reflected by a surface is called the **albedo** of that surface. Some typical albedos are given in Table 2.1, from which it is evident that polar ice-caps absorb only a relatively small proportion of the insolation. Waves and ripples significantly increase the albedo of water, but it is still generally less than that of most surfaces on land. The time of day will also affect the albedo (especially of water, ice or snow), because the shallower the angle of incidence of the solar radiation the greater the amount reflected.

Some of the radiation reflected back from the Earth's surface is absorbed in the atmosphere and warms it further. Also, because the surface is warmed by the radiation it has absorbed, it in turn radiates back infrared and longer (microwave) wavelengths.

Table 2.1 Some typical albedos.

Surface	Albedo (%)
Snow	up to 90
Desert sands	35
Vegetation	10–25
Bare soil or rock	10–20
Built-up areas	12–18
Calm water	2

QUESTION 2.1   At the same latitude, season and time of day, would you expect the atmosphere above snow-covered ground to be warmed more or less than that above a forest?

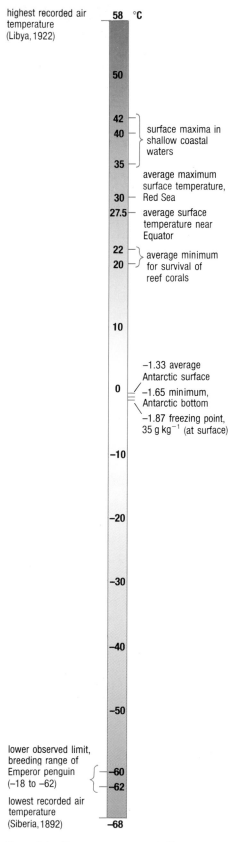

Figure 2.1 Temperature ranges in the sea (right) and on land (left).

highest recorded air temperature (Libya, 1922)

58 °C

50

42
40
} surface maxima in shallow coastal waters
35

average maximum surface temperature, Red Sea

30
27.5 — average surface temperature near Equator

22
20
} average minimum for survival of reef corals

10

−1.33 average Antarctic surface

0
−1.65 minimum, Antarctic bottom

−1.87 freezing point, 35 g kg⁻¹ (at surface)

−10

−20

−30

−40

−50

lower observed limit, breeding range of Emperor penguin (−18 to −62)

−60
−62

lowest recorded air temperature (Siberia, 1892)

−68

Atmospheric water vapour and carbon dioxide (CO_2) strongly absorb infrared wavelengths, and so the atmosphere acts as a blanket to keep in the heat. This has been called the **greenhouse effect** (probably incorrectly, because greenhouses mainly trap heat by preventing heated air from escaping by convection), and it is widely held that the increase in atmospheric CO_2 from the burning of fossil fuels such as coal and petroleum may be contributing to a general long-term warming of the atmosphere. CO_2 and water vapour are not the only 'greenhouse gases'. Other contributors to the effect are methane and nitrous oxide (see Table 6.2), and the artificially produced chlorofluorocarbons (CFCs, e.g. CCl_3F), the use of which as refrigerants and commercial aerosol propellants began to be phased out when it was recognized that they are also a major cause of ozone depletion in the upper atmosphere. The stratospheric 'ozone holes' observed above both Antarctic and Arctic regions have become well-publicized environmental issues because of the harmful biological consequences of increased ultraviolet radiation at the Earth's surface.

Diurnal (daily) variations of temperature on land are sometimes measurable in tens of degrees, but in the oceans they amount to no more than a few degrees, except in very shallow water (Figure 2.1).

QUESTION 2.2 With the help of Table 1.1 (p. 6), can you suggest three main reasons for this?

The answers to Question 2.2 account also for the contrasts displayed in Figure 2.1: the range of temperature in the oceans is about 40 °C (or about 30 °C if we exclude shallow and restricted seas); whereas the temperature range encountered on land is about three times greater. The temperature-buffering effect of the oceans (Questions 1.5 and 2.2) depends on the continuous exchange of heat and water between ocean and atmosphere, mainly by means of the hydrological cycle (Figure 1.3).

2.2 DISTRIBUTION OF SURFACE TEMPERATURES

The intensity of insolation depends primarily on the angle at which the Sun's rays strike the surface (Figure 2.2(a)), and the distribution of temperature over the surface of the Earth varies with latitude and season, because of the tilt of the Earth's axis with respect to its orbit round the Sun. Figure 2.2(b) shows that along the Equator maximum insolation occurs at the March and September **equinoxes**, when the noonday Sun is overhead. Insolation remains high in equatorial regions during the rest of the year. The noonday Sun is overhead along the Tropics of Cancer and Capricorn at the June and December **solstices** respectively, so temperate latitudes receive maximum and minimum insolation during their respective summer and winter seasons. There is insolation for only about half the year at the poles, which are wholly illuminated in summer and wholly dark in winter.

Until the advent of satellite technology, it was impossible to monitor seasonal changes of sea-surface temperature over wide areas. Satellite-mounted infrared sensors now make it possible to measure the change of sea-surface temperature on a global scale, both seasonally and from year to year (Figure 2.3). The sensitivity and precision of the sensors is of the order of ± 0.1 °C or better, and their accuracy is increasing all the time, as corrections are made for factors such as the state of the sea-surface (smooth or rough) and the amount of water vapour in the atmosphere (water vapour absorbs infrared radiation). Information such as that in Figure 2.3 can be obtained on a more or less continuous basis, and for many oceanographic purposes it is the *variation* in sea-surface temperature that is important, rather than the absolute values. It is essential to bear in mind that this is information about the sea-surface *only*. Satellite-based instruments cannot discover anything about the depth-related temperature structure of the oceans (see also Chapter 5).

Figure 2.2 (a) The angle of the Sun with respect to the Earth's surface determines both the distance the rays have to travel through the atmosphere and the surface area over which the energy will be spread.

(b) The four seasons (given for the Northern Hemisphere) related to the Earth's orbit around the Sun. The Earth's axis is tilted at approximately 23.4° to the plane of its elliptical orbit around the Sun, so at the northern summer and winter solstices the noonday Sun is overhead at the Tropics of Cancer (~23° N) and Capricorn (~23° S), respectively; at the spring and autumn equinoxes it is overhead at the Equator.

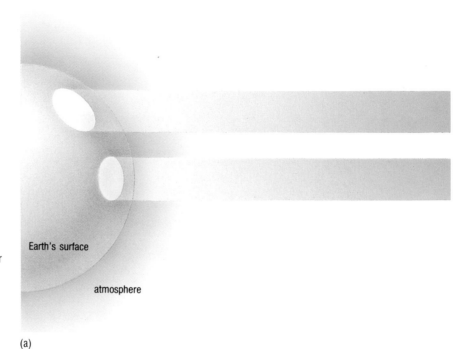

(a)

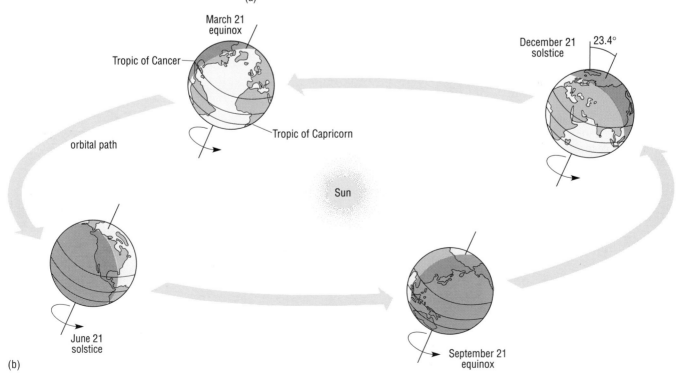

(b)

Figure 2.3   Daytime sea-surface temperature measurements from satellite-borne sensors. For the upper two pictures, temperatures below 0 °C are green and blue; higher temperatures are red and brown.

(*Top*): In January, the Northern Hemisphere experiences extreme cold. In Siberia and over most of Canada, temperatures approach −30 °C, while in Eastern Europe and the northern USA temperatures are below 0 °C. In the Southern Hemisphere it is summer, with mid-latitude temperatures ranging from 20 °C to 30 °C. On the eastern and western sides of the major oceans, the contours of equal temperature show deviations from their latitudinal (or *zonal*) patterns. Generally, in the sub-tropics of both hemispheres (*c.* 10° to 30°), the western sides of the oceans are warmer than their eastern counterparts, primarily due to ocean currents. The Gulf Stream can be seen moving along the eastern North American coast, then turning eastwards and transporting warm waters across the Atlantic to moderate the climate of north-west Europe.

(*Middle*): By July, areas of the Northern Hemisphere have warmed to 10–20 °C. Equatorial Africa and India are the hottest. In the Arctic, Greenland remains frozen, while Hudson Bay has thawed. In the Southern Hemisphere, ice has formed in the Weddell Sea, and Antarctica is even colder than the Arctic.

(*Bottom*): This shows temperature *differences* between January and July, and emphasizes the point made in Figure 2.1 and Question 2.2. The greatest warming and cooling has occurred over land (dark blue, brown). Marked seasonal changes of up to 30 °C are seen on land in both hemispheres. In contrast, the changes in ocean temperature rarely exceed 8–10 °C. The greatest deviations are in mid-latitudes, while tropical and equatorial regions are quite stable. In the Northern Hemisphere, mid-latitude changes in ocean temperature are strongly influenced by the position of continents. The continents divert the ocean currents and affect wind patterns. In the Southern Hemisphere, which has only half the land area of the Northern Hemisphere, changes are primarily due to seasonal variations of incoming solar radiation.

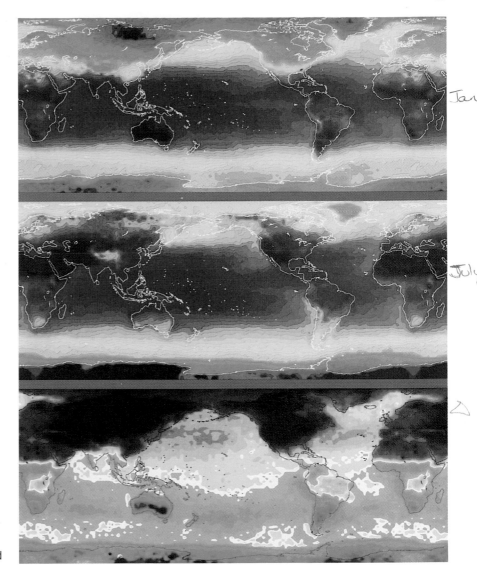

2.2.1   THE TRANSFER OF HEAT AND WATER ACROSS THE AIR–SEA INTERFACE

The surface temperature of the sea depends on the insolation and determines the amount of heat radiated back into the atmosphere: the warmer the surface, the more heat it radiates. Heat is also transferred across the surface of the sea by conduction and convection, and by the effects of evaporation.

Conduction and convection
If the sea-surface is warmer than the air directly above it, heat can be transferred from the sea to the air. On average, the sea-surface is warmer than the overlying air, so there is a net loss of heat from the sea by conduction. This loss is relatively unimportant in the total heat budget of the oceans, and its effect would be negligible were it not for convective mixing by wind, which removes the warmed air from just above the sea-surface.

Evaporation

Evaporation (the transfer of water to the atmosphere as water vapour) is the main mechanism by which the sea loses heat – about an order of magnitude more than is lost by conduction plus convective mixing.

The governing equation is:

(rate of loss of heat) = (latent heat of evaporation) × (rate of evaporation)   (2.1)

QUESTION 2.3 (a) Use Figure 1.3 to calculate an approximate value for the heat lost from the oceans by evaporation each day, using the value for latent heat of evaporation given in Table 1.1. In this context, the units in equation 2.1 are: $(J\,day^{-1}) = (J\,kg^{-1}) \times (kg\,day^{-1})$.

(b) Given that the Earth's surface and atmosphere receive about 9×10^{21} J from the Sun each day (70% of the incoming solar radiation, Section 2.1), would you say that evaporation from the oceans is a significant component in the Earth's heat budget?

(c) Under what conditions would you expect ocean water to *gain* heat by condensation?

Figure 2.4 (a) Diagrammatic representation of the successive stages of bubble collapse, for a 'typical' 1 mm diameter bubble.
(μm = micrometre (micron) = 10^{-6} m and ng = nanogram = 10^{-9} g.)

(b) Decrease in chloride content of rainwater with increasing distance inland from the coast.

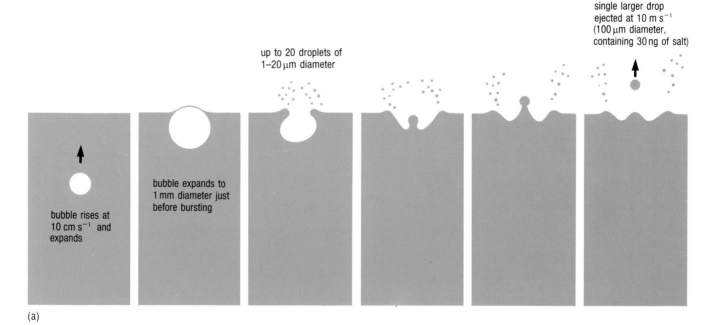

single larger drop ejected at 10 m s^{-1} (100 μm diameter, containing 30 ng of salt)

up to 20 droplets of 1–20 μm diameter

bubble expands to 1 mm diameter just before bursting

bubble rises at 10 cm s^{-1} and expands

(a)

(b)

Evaporation, condensation and precipitation are not the only mechanisms for transferring water across the interface between air and sea. As with all liquid bodies, the outer surface of the ocean is defined by intermolecular forces that cause a *surface tension*. The surface tension of seawater is less than that of freshwater, so seawater more readily breaks into froth or foam when disturbed by surface waves. High winds cause foaming and streaking of the surface layers, as well as entrapment of air bubbles.

Figure 2.4(a) shows what happens when air is injected into subsurface water under rough conditions, with breaking waves and white caps. Bubbles of trapped air rise to the surface and break, injecting droplets of various sizes into the atmosphere, along with any dissolved salts, gases and particulate matter that the water may contain. A large proportion of these constituents is soon returned to the Earth's surface by precipitation, as shown by the decrease in chloride content of rainwater with increasing distance inland from coasts (Figure 2.4(b)). The smallest droplets injected into the atmosphere are called **aerosols**, and they remove water, dissolved salts and organic matter from the surface of the oceans. Aerosols can be carried high above the Earth and dispersed throughout the atmosphere. When the water evaporates, the minute precipitated particles of salt and other substances act as nuclei for cloud and rain formation.

2.3 DISTRIBUTION OF TEMPERATURE WITH DEPTH

Measurement of temperature at the surface of the ocean, let alone below it, was not possible until the thermometer was invented in the early 17th century. The earliest temperature measurements were made on water samples collected in iron or canvas buckets from surface waters. It was realized that temperature decreased with depth, but accurate measurement of subsurface temperatures became possible only when thermometers protected against the water pressure and capable of recording *in situ* temperatures, were invented in the mid-19th century, shortly before the voyage of HMS *Challenger*. Temperature in the oceans is nowadays measured with thermistors, and continuous temperature recording – both vertical and lateral – is now a routine oceanographic procedure.

Figure 2.5 shows that most solar energy is absorbed within a few metres of the ocean surface, directly heating the surface waters and providing the energy for photosynthesis by marine plants and algae.

Why is the colour of the light below the surface of the sea predominantly blue–green? Which wavelengths are the first to be absorbed? What proportion of the total incident energy reaches a depth of 100 m?

Shorter wavelengths, i.e. those near the blue end of the visible spectrum, penetrate deeper than longer wavelengths. Infrared radiation is the first to be absorbed, followed by red, and so on. The total energy received at a given depth is represented by the area beneath the relevant curve on Figure 2.5. Comparison of the areas beneath the curves for 100 m and the water surface suggests that only about one-fiftieth of the incident energy penetrates to 100 m. All of the infrared radiation is absorbed within about a metre of the surface, and nearly half of the total incident solar energy is absorbed within 10 cm of the surface. Penetration will also depend on the

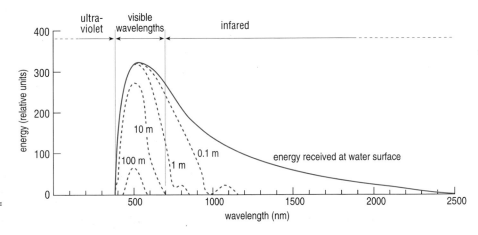

Figure 2.5 A simplified energy–wavelength spectrum of solar radiation at the surface of the ocean and at various depths. (nm = nanometre = 10^{-9} m.)

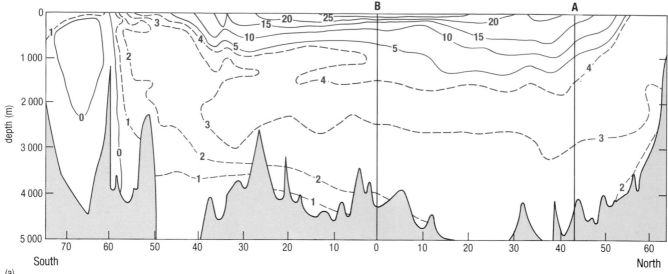

(a)

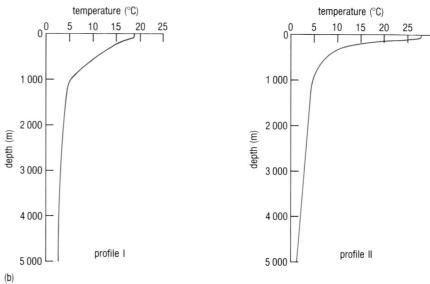

(b)

Figure 2.6 (a) A vertical section showing the mean distribution of temperature in the western Atlantic Ocean to illustrate that the range of temperature in surface layers is much greater than that in the main body of ocean water below 1 000 m. This general pattern is typical of all ocean basins (although the detail will vary from ocean to ocean). Note the great vertical exaggeration. Contours of equal temperature are called **isotherms**. Broken line isotherms at 1 °C interval; solid lines at 5 °C interval. The vertical lines A and B relate to Figure 2.6(b) and are for use with Question 2.4.

(b) Temperature profiles along A and B in (a), for use with Question 2.4.

clarity or transparency of the water, which in turn depends on the amount of suspended matter in it (see also Chapter 5).

If the thermal energy from solar radiation is largely absorbed by the surface layers, how can it be carried deeper?

Conduction by itself is extremely slow, so only a small proportion of heat is transferred downwards by this process. The main mechanism is turbulent mixing by winds and waves, which establishes a **mixed surface layer** (often called simply the mixed layer) that can be as thick as 200–300 m or even more at mid-latitudes in the open oceans in winter, and as little as 10 m thick or less in sheltered coastal waters in summer.

Between 200–300 m and 1 000 m depth, the temperature declines rapidly throughout much of the ocean. This region of steep temperature gradient is known as the **permanent thermocline**, beneath which, from about 1 000 m to the ocean floor, there is virtually no seasonal variation and temperature decreases gradually to between about 0 °C and 3 °C (Figure 2.7(a–c)).

This narrow range is maintained throughout the deep oceans, both geographically and seasonally, because it is determined by the temperature of cold, dense water that sinks from the polar regions and flows towards the Equator (see Section 4.1).

QUESTION 2.4   Figure 2.6(a) is a vertical section illustrating the range of temperatures encountered in the oceans, and Figure 2.6(b) shows temperature profiles along lines A and B in Figure 2.6(a).

(a)   Match the profiles I and II in Figure 2.6(b) with vertical lines A and B in Figure 2.6(a).

(b)   What can you say about the vertical distribution of temperature at high latitudes (above about 60° N and 60° S)?

Important notes

Sections and profiles: A vertical *section* is an imaginary slice through a part of the ocean, to show both vertical and horizontal distribution (commonly represented by contours) of some property (temperature, salinity, density, and so on); e.g. Figure 2.6(a). A vertical *profile* is a graph showing how some property (temperature, salinity, etc.) varies with depth at a single location in the ocean; e.g. Figure 2.6(b).

Gradients on profiles: A near-vertical part of a temperature profile means that there is *little* change of temperature with depth (e.g. lower parts of profiles in Figure 2.6(b)). A near-horizontal part of a temperature profile means a *large* change of temperature with depth (e.g. upper parts of profiles in Figure 2.6(b)). So, when you read or hear of a 'steep temperature gradient' or 'steep thermocline', that is where the profile is near-horizontal and the rate of change of temperature with depth is greatest. Exactly the same applies to profiles for any other property (salinity, density, etc.).

Above the permanent thermocline, the distribution of temperature with depth shows seasonal variations, especially in mid-latitudes. During the winter, when surface temperatures are low and conditions at the surface are rough, the mixed surface layer may extend to the permanent thermocline; i.e. the temperature profile can be effectively vertical through the top 200–300 m or more. In summer, as surface temperatures rise and conditions at the surface are less rough, a **seasonal thermocline** often develops above the permanent thermocline, as shown in the generalized profile of Figure 2.7(a).

Seasonal thermoclines start to form in spring and reach their maximum development (i.e. with greatest rate of change of temperature with depth or steepest temperature gradients) in the summer. They develop at depths of a few tens of metres, with a thin mixed layer above (Figure 2.7(a)). Winter cooling and strong winds progressively increase the depth of seasonal thermoclines and reduce the temperature gradient along them; eventually the mixed layer reaches its full thickness of 200–300 m (see Figure 2.7(d)). In low latitudes, there is no winter cooling, so the 'seasonal thermocline' becomes 'permanent' and merges with the permanent thermocline at depths of 100–150 m (Figure 2.7(b)). At high latitudes greater than about 60° there is no permanent thermocline (Figures 2.6 and 2.7(c)), though seasonal thermoclines can still develop in summer.

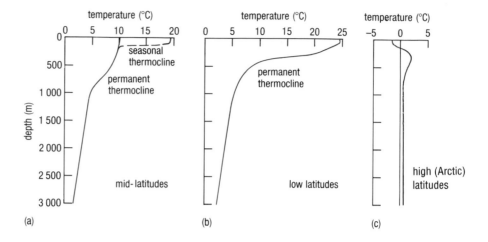

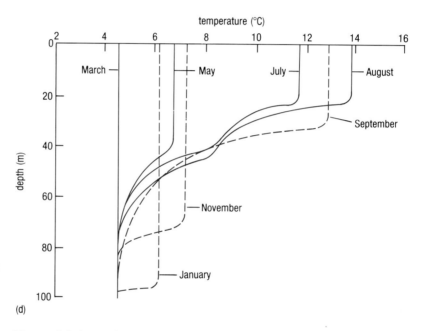

Figure 2.7 (a)–(c) Typical mean temperature profiles for different latitude belts in the open oceans. Note that the otherwise vertical profile (c) for high (Arctic) latitudes shows a layer of colder water at the surface overlying slightly warmer water at *c.* 200 m, depth. (d) A succession of temperature profiles to show the growth (solid lines) and decay (broken lines) of a seasonal thermocline in the Northern Hemisphere. Note the very different scales compared to (a)–(c).

Figure 2.8 (opposite) gives an idea of the pattern of temperature change with depth and season at mid-latitude. The annual range of nearly 10 °C at the surface declines to only about 3–4 °C at 100 m depth.

QUESTION 2.5 Given that changes of temperature resulting directly from seasonal variations of incident radiation can no longer be detected below about 200 m, approximately where would you place the curve for 200 m on Figure 2.8, and what form would you expect it to have?

Diurnal thermoclines can form anywhere, provided there is enough heating during the day, though they extend only to depths of about 10–15 m, and temperature differences across them do not normally exceed 1–2 °C.

In summary, and ignoring seasonal and diurnal variations, the permanent thermocline allows the oceans as a whole to be divided into three principal layers, shown schematically in Figure 2.9. The thickness of both the upper warm layer and permanent thermocline is less at low latitudes than at mid-latitudes because at low latitudes winds are generally weaker and seasonal temperature contrasts are less.

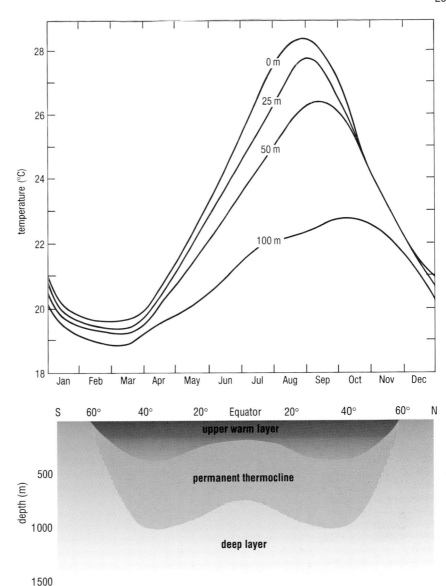

Figure 2.8   Annual variations of water temperature at different depths in the ocean off the south coast of Japan (about 25° N).

Figure 2.9   Generalized and schematic cross-section, showing the main thermal layers of the oceans. The base of the upper warm layer approximates to the 10 °C isotherm. Seasonal variations are largely confined to this layer (including development of seasonal and diurnal thermoclines during summer in mid-latitudes).

2.4   ENERGY FROM THE THERMOCLINE – A BRIEF DIGRESSION

The permanent thermocline is found nearly everywhere in the oceans (Figure 2.9) and in low latitudes the temperature difference across it is of the order of 20 °C, and sometimes more (Figures 2.6 and 2.7). The problems of tapping energy from this temperature gradient in ocean waters are mainly those of scale. The principle of Ocean Thermal Energy Conversion (**OTEC**) is exactly the same as that used in refrigerators, air conditioners and heat pumps.

The original concept was to pump warm surface water at about 25 °C into heat exchangers to vaporize a volatile liquid (such as ammonia) which would expand to drive turbines to generate electricity. At the same time, cold water at about 4 °C from below the thermocline would be pumped up to condense the vapour in separate heat exchangers, allowing the cycle to

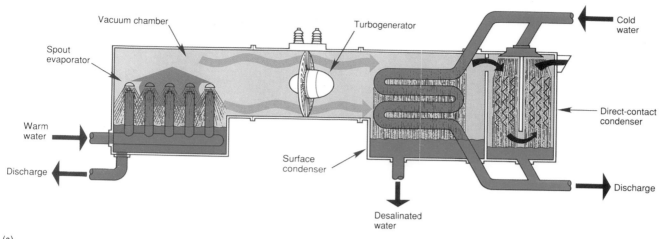

(a)

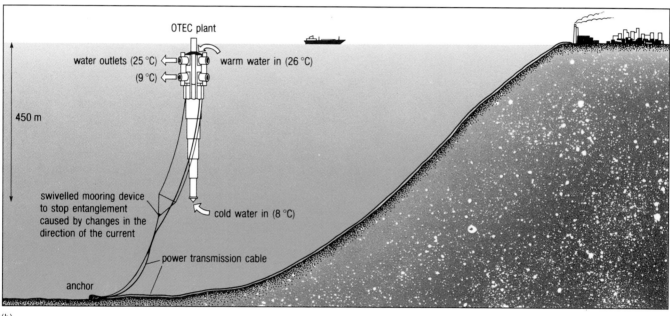

(b)

Figure 2.10 (a) Schematic diagram of an OTEC plant on Hawaii. Warm seawater is pumped into a vacuum chamber, where it boils and evaporates after passing through a specially designed spout evaporator, producing steam. The steam is used to drive a turbo-generator to generate electrical power. The steam then passes to a condenser chamber, where cold seawater is used to condense the steam back into water. The cold seawater (which is nutrient-rich, see Section 6.1.2) can then be used for mariculture (fish or shellfish farming), while the desalinated water (condensed steam) can be used for freshwater and/or irrigation.

(b) Artist's impression of an OTEC plant which would use the difference in temperature between the surface of the ocean and deep waters to obtain electrical energy. As much as 160 000 kW might be generated by such a plant.

start again. In some modern plants (Figure 2.10(a)), the warm seawater itself is vaporized under a vacuum and the resulting steam used to drive turbines.

Such plants are best installed in lower latitudes, where the thermal contrast between surface and deep water is greatest and seasonal changes are least. The Japanese and Americans have advanced furthest with this technology and have built some small plants generating 50 to 100 kW. At this scale, the most likely beneficiaries of the technology are small islands in the South Pacific.

For larger power stations (hundreds of megawatts or more), huge installations would be required, comparable in size to oil production platforms on continental shelves (Figure 2.10(b)). That is because the temperature differential between surface and deep water is only about 25–30 °C at best, so the 'energy density' of the vapour driving the turbines is low compared with that in conventional steam turbines for example, where the temperature differential is several hundred degrees. What is more, around two-thirds of the power output is needed to drive the pumps, so overall efficiency of OTEC plants is unlikely to be better than 5–10%.

2.5 TEMPERATURE DISTRIBUTION AND WATER MOVEMENT

Sections and profiles such as those in Figures 2.6 and 2.9 represent temperatures averaged over periods of months or years. We know that large seasonal changes of temperature occur in the surface layers (e.g. Figure 2.8), and there can be small fluctuations with time, even in the deep oceans.

It is most important, however, not to gain the impression from such time-averaged temperature sections and profiles that the waters of the oceans are static. Far from it. It is essential always to bear in mind that while the locations of mean isotherms along such sections do not change significantly even on time-scales of decades, the structure is maintained dynamically. Any given parcel of water can travel a distance equivalent to a global circumnavigation in a few years; but the average temperature structure at a particular location remains much the same. In other words, the temperature (and, as you will see, the salinity) at any particular location and depth – at least below the mixed surface layer – changes very little from year to year, even though the actual *water* at that location and depth is changing all the time.

Figure 2.11 The world-wide average pattern of oceanic surface currents for November to April. Cool currents are shown by dashed arrows, warm currents by solid arrows. From May to October, there are significant differences in the Indian and western Pacific Oceans due to reversals of winds and currents during the monsoons.

We have seen that the distribution of oceanic surface temperatures is in part the direct result of insolation, and that it varies seasonally (Figure 2.3 and related text). Just as important are the processes of horizontal **advection** (horizontal movements) that transport warm water to cooler regions and *vice versa*. Figure 2.11 is a time-averaged map of the major surface current systems, and Figure 2.12 shows the average global distribution of surface isotherms for (a) northern winter/southern summer and (b) northern summer/southern winter.

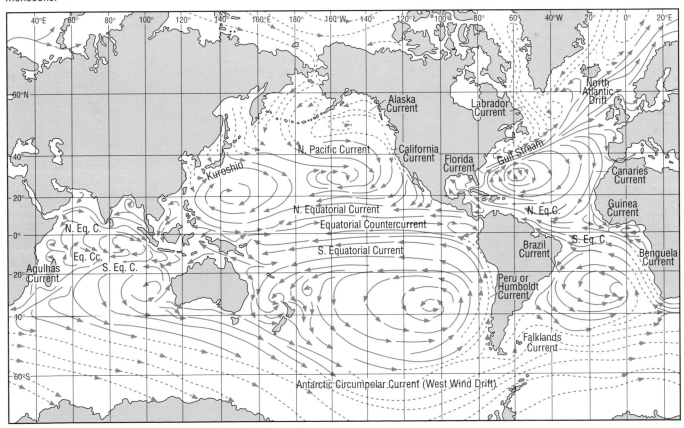

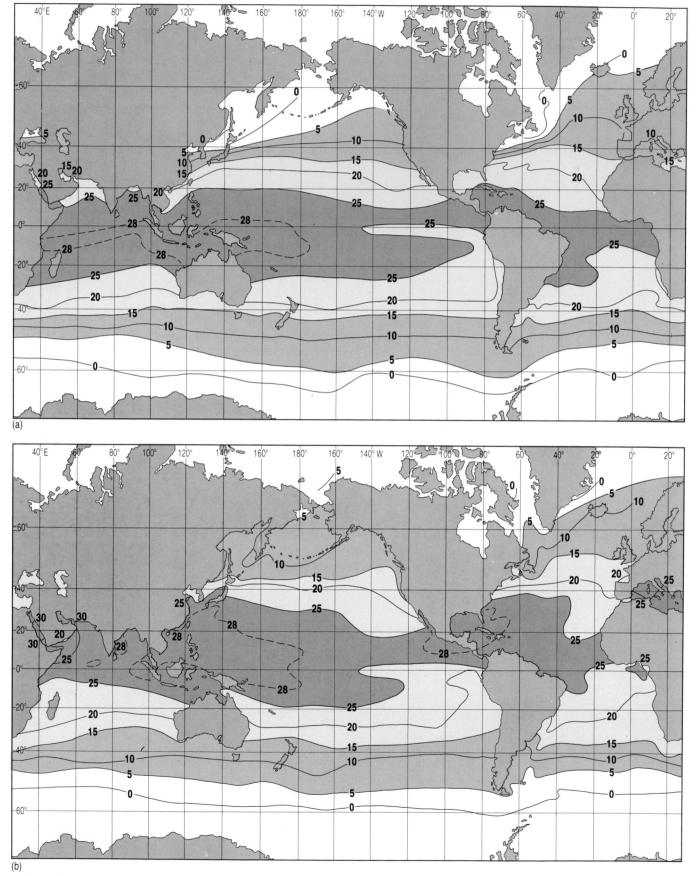

Figure 2.12 The global distribution of sea-surface temperature (°C) (a) in February, (b) in August.

Compare Figures 2.11 and 2.12. Can you detect the influence of major surface currents on the distribution of isotherms on each side of the Atlantic and Pacific Oceans?

There is a general poleward displacement of isotherms on the western sides of these ocean basins and an equatorwards displacement on the eastern sides (also seen in Figure 2.3). Poleward-flowing currents carry warm water from low to high latitudes – and the effect of the Gulf Stream in carrying relatively warm water across the Atlantic can be clearly seen. Currents flowing towards the Equator (e.g. Canaries Current, Benguela Current) carry cool water from high to low latitudes, resulting in lower average surface temperatures on eastern margins of the ocean basins.

Vertical distribution of temperature in the oceans (e.g. Figure 2.6) is controlled by density-driven vertical water movements (see Section 4.1). Figure 2.13 illustrates in a generalized way how the convective sinking of cold dense water in polar regions drives the circulation in the deep oceans.

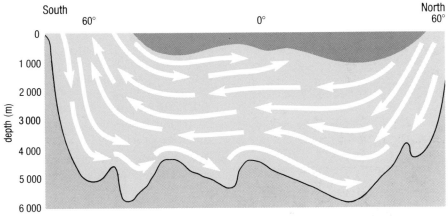

Figure 2.13 Diagrammatic section to illustrate the general form of the deep circulation in the Atlantic Ocean, driven by cold dense water sinking in high latitudes (the base of the pink shaded region at the top approximates to the 10 °C isotherm, cf. 'upper warm layer' on Figure 2.9). Deep currents also circulate through the other major ocean basins.

2.6 SUMMARY OF CHAPTER 2

1 The Earth's surface temperature is mainly determined by the amount of solar radiation it receives. On average, about 70% of incoming solar radiation reaches the surface, directly or indirectly. The proportion varies with latitude, season and time of day, and the amount absorbed depends on the albedo of the surface. The oceans have a large thermal capacity because of the high specific and latent heat of water, and they act as a temperature buffer for the surface of the Earth as a whole. Annual insolation is greatest in low latitudes and least at the poles, mainly because of the angle that the Sun's rays make with the Earth's surface: the higher the latitude, the lower the angle.

2 Conduction, convection and especially evaporation/precipitation are the principal means by which heat and water are exchanged across the air–sea interface. The oceanic evaporation/precipitation cycle contributes about one-quarter of the global heat budget. Aerosol production at the sea-surface is another important mechanism for the transfer of water (and salts) into the atmosphere.

3   Solar radiation penetrates no more than a few hundred metres into the oceans, and most is absorbed within the topmost 10 m. Downward transfer of heat is mainly by mixing, as conduction is very slow (water is a very poor conductor of heat). Mixing by winds, waves and currents produces a mixed surface layer which can be 200–300 m thick or more in winter in mid-latitudes. Below this lies the permanent thermocline, across which temperature declines to about 5 °C, and below which temperature decreases gradually to the bottom (typically between 0 °C and 3 °C). In mid-latitudes, a seasonal thermocline can develop during summer, above the permanent thermocline. There may also be diurnal thermoclines, at depths of 10–15 m.

4   The temperature difference across the permanent thermocline can be utilized to generate electricity, using the principles upon which the domestic refrigerator is based. The main problem in this application is that of scale.

5   The long-term stability of the distribution of temperature within the ocean means that sections and profiles of average temperature do not change significantly from year to year. This stable thermal structure is maintained by the continuous three-dimensional motion of the global system of surface and deep currents.

Now try the following questions to consolidate your understanding of this Chapter.

QUESTION 2.6   (a)   Explain why water at 4 °C can be overlain by colder water in a freshwater lake in a gravitationally stable situation. Could such a situation develop in the oceans?

(b)   Explain why a temperature profile from a freshwater lake will not show temperatures decreasing with depth to values of less than 4 °C.

QUESTION 2.7   The broad thermal structure of the oceans allows us to recognize three main layers. Name them and summarize their characteristics.

CHAPTER 3 SALINITY IN THE OCEANS

The average concentration of dissolved salts in the oceans – the **salinity** (*S*) – is about 3.5 per cent (%) by weight. Until the 1980s, salinity values were expressed in parts per thousand, or per mil, for which the symbol is ‰; the average salinity quoted above is therefore 35‰. It has become standard practice to dispense with the symbol because salinity is now defined in terms of a ratio, as explained in Section 3.3.3, and we shall normally dispense with this symbol hereafter and use only numbers when presenting salinity values. Table 3.1 lists the 11 major ions that together make up 99.9% of the dissolved constituents of seawater. The symbol ‰ appears in Table 3.1 only to remind you that in practice the numbers represent concentrations in parts per thousand by weight (i.e. either grams per kilogram ($g\,kg^{-1}$) or grams per litre ($g\,l^{-1}$), because for most purposes it can be assumed that one litre of seawater weighs one kilogram).

In surface waters of the open oceans, salinity ranges from 33 to 37, but when shelf seas and local conditions are taken into account, the range can be as wide as 28–40 or more. **Brackish water** has a salinity of less than about 25, while **hypersaline** water has a salinity greater than about 40.

Table 3.1 Average concentrations of the major ions in seawater, in parts per thousand by weight ($g\,kg^{-1}$ or $g\,l^{-1}$).

Ion	‰ by weight	
Chloride, Cl^-	18.980	
Sulphate, SO_4^{2-}	2.649	
Bicarbonate,* HCO_3^-	0.140	Negative ions (**anions**) total $= 21.861$‰
Bromide, Br^-	0.065	
Borate, $H_2BO_3^-$	0.026	
Fluoride, F^-	0.001	
Sodium, Na^+	10.556	
Magnesium, Mg^{2+}	1.272	
Calcium, Ca^{2+}	0.400	Positive ions (**cations**) total $= 12.621$‰
Potassium, K^+	0.380	
Strontium, Sr^{2+}	0.013	
Overall total salinity	34.482‰	

* Includes carbonate, CO_3^{2-}.

Note: You will find tables of average concentrations of the dissolved constituents of seawater elsewhere, both in this Series and in other literature. They differ in detail from Table 3.1 because different authorities use different sources and methods to compile their averages.

Table 3.2 Approximate average percentages by weight of the ten most abundant elements (other than oxygen) in the Earth's crust.

Element	% by weight
Silicon, Si	28.2
Aluminium, Al	8.2
Iron, Fe	5.6
Calcium, Ca	4.2
Sodium, Na	2.4
Potassium, K	2.4
Magnesium, Mg	2.0
Titanium, Ti	0.6
Manganese, Mn	0.1
Phosphorus, P	0.1

QUESTION 3.1 In Table 3.1, the proportion by weight of negative ions (anions) greatly exceeds that of positive ions (cations). So why does seawater not carry a net negative charge?

Compare Table 3.1 with Table 3.2, which presents an approximate average elemental composition of crustal rocks: there are some obvious contrasts. These are particularly striking when you realize that the operation of the hydrological cycle provides most of the dissolved constituents in seawater.

However, since the late 1970s oceanographers have recognized another important contribution to seawater composition: **hydrothermal circulation** at ocean ridge crests.

QUESTION 3.2 How many of the most abundant elements in common crustal rocks (Table 3.2) can you find in Table 3.1, and which are they?

The three most abundant elements in Table 3.2 – let alone others – do not appear at all in Table 3.1. The reason lies in the degree of solubility and the chemical behaviour of different elements when rocks are weathered and the resulting products are carried away by rivers to the sea. Many of the commonest elements in rocks, such as silicon, aluminium and iron, are not very soluble, and so they are transported and deposited mainly in solid particles of sand and clay. Others such as sodium, calcium and potassium are relatively soluble and are transported mainly in solution. Hydrothermal solutions associated with sea-floor spreading supply some elements to the seawater solution (e.g. calcium, silicon, manganese) and remove others from it (e.g. magnesium, sulphur). The relative amounts of dissolved constituents within the oceans are controlled by complex chemical and biological reactions in seawater, as you will see in Chapter 6.

3.1 CONSTANCY OF COMPOSITION

The **constancy of composition of seawater** is an important concept in oceanography. For most of the major ions in Table 3.1, the following generalization applies:

> The *concentrations* of the major dissolved ions can vary from place to place in the oceans, but their *relative proportions* remain virtually constant.

In other words, the total salinity can change, but the ratio of the concentration of any particular major ion to the total remains virtually constant, and so do the ratios of the concentrations of individual major ions to one another.

QUESTION 3.3 (a) What is the ratio of potassium concentration to total salinity in Table 3.1?

(b) What would the potassium concentration be if the salinity in Table 3.1 (i) rose to 36, (ii) fell to 33?

(c) What is the ratio of the concentrations of K^+ to Cl^- shown in Table 3.1? What would it be in each of cases (i) and (ii) of part (b)?

(d) How might these changes of salinity come about?

The way salinity varies throughout the oceans depends almost entirely on the balance between evaporation and precipitation and the extent of mixing between surface and deeper waters. In general, changes of salinity have no effect on the relative proportions of the major ions – their concentrations all change in the same proportion, i.e. their ionic ratios remain constant.

Exceptions to this generalization are relatively small variations in the ratios of calcium and bicarbonate,[*] because of their involvement in biological

---

[*] The term 'bicarbonate' for HCO_3^- is nowadays sometimes replaced by the term 'hydrogen carbonate'. However, we shall use 'bicarbonate' as it is still widely used in the literature.

processes (see Section 6.1.2): the ratios of Ca^{2+} and HCO_3^- to total salinity are respectively about 0.5% and about 10–20% greater in deep than in surface waters.

3.1.1 CHANGES DUE TO LOCAL CONDITIONS

In some marine environments, conditions are such that ionic ratios show large departures from normal. Such regions include:

1 Enclosed seas, estuaries and other regions where there is a substantial inflow of river water, which not only contains significantly less total dissolved salts than seawater, but also has very different ionic ratios (see Section 6.2.1).

2 Basins, fjords and other regions where the bottom circulation is severely restricted, e.g. by the presence of a sill (a subsurface barrier) at the mouth of the basin, preventing free communication between the bottom water and the oxygenated oceanic water outside. In such cases the bacterial breakdown (oxidation) of organic matter in the bottom water leads to depletion of dissolved oxygen, which may be severe enough to result in total depletion, conditions that are described as **anoxic** or **anaerobic**. Sulphate anions are then used as an alternative source of oxygen by the micro-organisms.

3 Extensive areas of warm, shallow water, such as the Bahama Banks, which are characterized by very active chemical and/or biological precipitation of calcium carbonate, leading to significant changes in the ratio of Ca^{2+} to total salinity.

4 Regions of sea-floor spreading and active submarine volcanism, where heated seawater circulates through cracks and fissures in the oceanic crust. Ionic ratios in hydrothermal solutions are very different from those of normal seawater, and the resulting mixtures with seawater have atypical major element : salinity ratios.

5 Within sea-floor sediments where interstitial or pore waters participate in a wide variety of reactions with the sediment particles during compaction, after the sediments have been deposited. Such reactions come under the general heading of **diagenesis** and can lead to considerable changes in ionic ratios.

QUESTION 3.4   As Table 3.1 shows, sulphur is present in seawater mainly as SO_4^{2-}, and in practice it is measured in this form. Would you expect the ratio of SO_4^{2-} to total salinity to be higher or lower in anoxic basins (item 2 in the above list) than in open ocean waters?

3.1.2 SALTS FROM SEAWATER

As seawater evaporates, the least soluble salts reach saturation first, so the sequence of precipitation is in the order of increasing solubility, not of abundance. The sequence is shown in Figure 3.1, along with the relative proportions of the precipitated salts. The first to be precipitated is calcium carbonate ($CaCO_3$), which forms only a small proportion of the salts because of the relatively low abundance of bicarbonate (and carbonate) ions (Table 3.1).

Calcium sulphate is precipitated either as anhydrite ($CaSO_4$) or as gypsum ($CaSO_4.2H_2O$), depending upon the conditions. Sodium chloride (halite, NaCl) is the most abundant salt, and the residual brine contains the

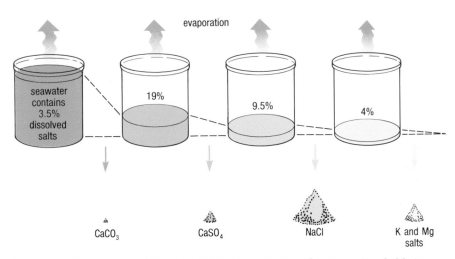

Figure 3.1 The succession of salts precipitated from seawater. On evaporation, $CaCO_3$ is precipitated first. When evaporation has reduced the volume to 19% of the original amount, $CaSO_4$ begins to precipitate; at 9.5% of the original volume, NaCl starts to precipitate, and so on. The volumes of the piles represent the relative proportions of the precipitated salts.

chlorides of potassium and magnesium, which are the most soluble and therefore the last to be precipitated.

Virtually every coastal nation has at some time produced sea salt commercially and about 60 countries still do, either by industrial processes or by solar evaporation (Figure 3.2). Some 40 million tonnes of sodium chloride are extracted from seawater each year world-wide, some for human consumption but most of it for manufacturing chemicals. Magnesium hydroxide is chemically precipitated from seawater and used to produce around 600 000 tonnes of magnesium and its compounds annually. Bromine is released by electrolysis as a gas and then condensed to liquid – annual production is about 30 000 tonnes. A method for extracting lithium (Li) from seawater was developed in the late 1980s.

Most elements dissolved in seawater occur in minute concentrations (see Section 6.1.1), but the total volume of seawater is so huge that tonnages are enormous and efforts to extract valuable elements such as gold and uranium have been made many times; but no technique has so far proved economic.

Figure 3.2 (a) Raking sea salt out of the residual brine in solar evaporation pans near Aveiro in Portugal.

(b) Carrying salt from the piles to lorries or boats for transport to the purification and processing centres.

(a)

(b)

3.2 VARIATIONS IN SALINITY

Distributions of temperature and salinity together provide oceanographers with information that enables them to trace the three-dimensional pattern of oceanic circulation. This Section describes how salinity varies vertically and horizontally in the oceans. As with temperature distribution, the maps, sections and profiles illustrate a long-term stability of salinity distribution which is maintained dynamically. As you read, bear in mind that salinity at any particular location may hardly change from year to year, but the water at that location is changing all the time (cf. Section 2.5).

3.2.1 DISTRIBUTION OF SALINITY WITH DEPTH

Figure 3.3 shows a vertical section illustrating the relatively restricted range of salinity encountered in the main body of the oceans. Salinity is determined by the balance between precipitation and evaporation at the surface (cf. Question 3.3(d)). The influence of surface fluctuations is generally small below about 1 000 m, where salinities are mostly between about 34.5 and 35 at all latitudes.

Figure 3.3 (a) A vertical section showing the distribution of salinity in the western Atlantic Ocean, which illustrates that the range of salinity in surface layers is much greater than in the main body of ocean water below 1 000 m. This general pattern is typical of all ocean basins, although the details vary from ocean to ocean. Note the great vertical exaggeration. Lines joining points of equal salinity are called **isohalines**. Broken lines 0.1 and 0.2 interval; solid lines 0.5 interval. The vertical lines A and B relate to (b) and are for use with Question 3.5.

(b) Salinity profiles along lines A and B in (a), for use with Question 3.5.

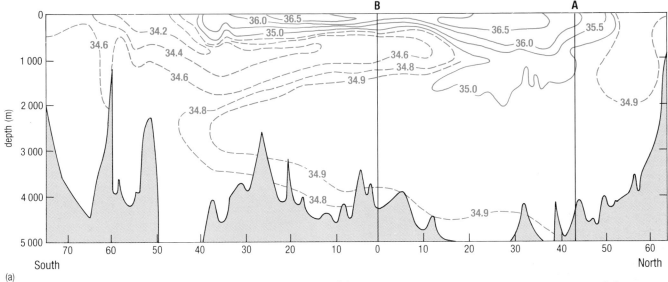

(a)

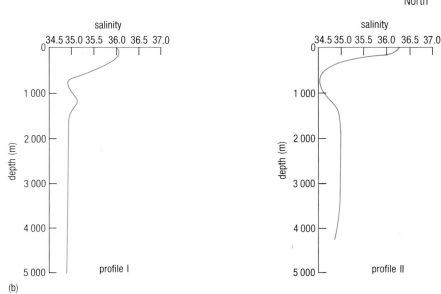

(b)

Zones where salinity decreases with depth are typically found at low and middle latitudes, between the mixed surface layer and the top of the deep layer, in which the salinity is roughly constant. These zones are known as **haloclines**; the term applies also to zones where salinity *increases* with depth (whereas in thermoclines temperature almost invariably decreases with depth).

QUESTION 3.5 (a) Which salinity profile in Figure 3.3(b) corresponds to which vertical line in Figure 3.3(a)? How do the depth ranges of the haloclines compare with those of the thermoclines shown earlier in Figure 2.6?

(b) Which of the two haloclines in Figure 3.3(b) shows the greater rate of decrease of salinity with depth?

(c) Would you expect salinity profiles at high latitudes to resemble those in Figure 3.3(b)?

3.2.2 DISTRIBUTION OF SURFACE SALINITY

The salinity of the surface waters of the oceans is at a maximum in tropical and sub-tropical latitudes, where evaporation exceeds precipitation. These regions correspond to the hot barren deserts that exist in similar latitudes on land. Salinities decrease both towards higher latitudes and towards the Equator (Figure 3.4). Local modifications are superimposed on this regional pattern, particularly near land masses. Surface salinity may be reduced by an influx of freshwater at the mouths of large rivers, and by melting ice and snow at high latitudes. On the other hand, surface salinities tend to be high in lagoons and other partly enclosed shallow marine basins at low latitudes, where evaporation is high and the inflow of water from adjacent land areas is limited.

Figure 3.4 (a) The approximate positions of mean annual surface isohalines.

(b) Average values of surface salinity, S (black line), and the difference between average annual evaporation and precipitation ($E - P$) (blue line), plotted against latitude.

QUESTION 3.6 Examine the map (a) and the curves (b) in Figure 3.4, and explain the maxima and minima on the curves; then account for the salinity minimum in equatorial latitudes.

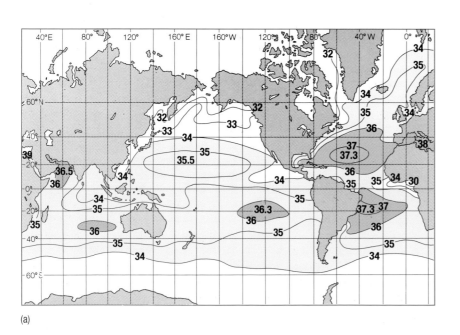

(a)

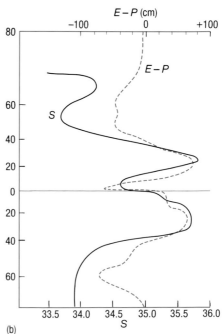

(b)

3.3 THE MEASUREMENT OF SALINITY

Early attempts to determine the chemical composition of seawater were hampered by the low sensitivity of analytical techniques. It was not until the early 19th century that any order became apparent in the data and the constancy of composition of seawater was first recognized from the few analyses available. During the cruise of HMS *Challenger* from 1872 to 1876, 77 water samples were collected from various depths in nearly all the major oceans and seas, and analysed for the elements chlorine, sodium, magnesium, sulphur, calcium, potassium and bromine. The method used for each element was rigorously tested on synthetic samples, thus giving a check on the reliability of the technique.

Since the 19th century, a large number of investigations have been carried out into the ratio of single constituents to salinity. During the mid-1960s, scientists from the British National Institute of Oceanography (now the Institute of Oceanographic Sciences) and the University of Liverpool analysed more than 100 samples for all the major constituents. In the 1970s, the GEOSECS programme (GEochemical Ocean SECtionS), based in the USA, collected systematic chemical data for all the oceans, using the most accurate analytical techniques then available and (more importantly) sampling procedures that minimized contamination. The huge amount of data collected is still being interpreted, and the GEOSECS measurements are now being supplemented, updated and gradually superseded, as more samples are collected for new research programmes and as analytical methods become more refined.

3.3.1 CHEMICAL METHODS OF SALINITY MEASUREMENT

The most obvious way of measuring salinity is to take a known amount of seawater, evaporate it to dryness and then weigh the remaining salts (gravimetric determination). Although simple in theory, such a method gives unreliable results, for a number of reasons. The residue left after evaporation is a complex mixture of salts, together with some water chemically bound to the solids, plus a small amount of organic material. The amount of water left behind can obviously be decreased by thorough drying of the residual salts at elevated temperatures, but this leads to other problems such as: (i) decomposition of some of the salts (e.g. loss of water and gaseous HCl from hydrous $MgCl_2$ crystals); (ii) vaporization and decomposition of the organic matter; and (iii) expulsion of CO_2 gas from carbonate salts. The weight of solid material left behind after evaporation (and hence the measured salinity) thus depends on the conditions employed to drive off the water. Marine chemists in the 19th century were well aware of this in their attempts to measure salinity gravimetrically, and devised procedures that gave reasonably reproducible results.

None the less, gravimetric determination of salinity is both difficult and tedious, so other methods were investigated. As you have read in Section 3.1, the concentrations of many major dissolved constituents of seawater bear a constant ratio to the total dissolved salt concentration, so the concentrations of one or more major constituents can be used to deduce the total salinity, S. The easiest constituents to measure are the halides (chloride + bromide + iodide), and this led ultimately to the empirical relationship:

$$S = 1.806\,55\ Cl \tag{3.1}$$

where Cl is the **chlorinity** of the sample, defined as the concentration of chloride in seawater (in parts per thousand) assuming that the bromide and iodide have been replaced by chloride.

Chlorinity was measured by titration and the salinity determined by substitution in equation 3.1 or tables derived from it. This method was used for determining virtually all salinities from the turn of the century until the mid-1960s. It is rarely used today, having been almost entirely superseded by measurements of electrical conductivity.

3.3.2 PHYSICAL METHODS OF SALINITY MEASUREMENT

Pure water is a poor conductor of electricity. However, the presence of ions in water enables it to carry an electric current. In the 1930s, it was established that the electrical **conductivity** of seawater is proportional to its salinity. Conductivity is inversely proportional to resistivity, and for many decades conductivity salinometers were based on simple electrical bridge circuits, using 'standard seawater' of known salinity (close to 35) for calibration.

Conductivity is also affected by temperature, however, which can lead to appreciable errors. Ideally, physical oceanographers require salinity measurements to be accurate to ± 0.001, requiring conductivity to be measured to 1 part in 40 000. A change of this magnitude can be induced by a temperature change of only 0.001 °C, so careful control of temperature is essential.

In the past, precision thermostatting was used to maintain both sample and standard seawater at constant temperature, but the equipment was bulky and measurements took a long time because samples had to be heated or cooled to working temperature before measurement could begin. Such problems have now been largely circumvented, and modern salinometers are compact and rapid in operation, and can measure salinity to ± 0.003 or better. Conductivity sensors have been incorporated into *in situ* temperature–salinity instruments for use in shallow waters, and into conductivity–temperature–depth (CTD) probes for use in the deep oceans.

3.3.3 THE FORMAL DEFINITION OF SALINITY

Since the mid-1960s, the definition of salinity has been based (by international agreement) on empirically determined and rather complicated-looking formulations involving a conductivity standard.

The salinity of a sample of seawater is now measured in terms of the conductivity ratio, R, which is defined by:

$$R = \frac{\text{conductivity of seawater sample}}{\text{conductivity of standard KCl solution}} \tag{3.2}$$

the concentration of the standard KCl solution being 32.435 6 g kg^{-1}.

Salinity is related to the conductivity ratio at 15° C and 1 atmosphere pressure (R_{15}) by the following equation:

$$S = 0.008\,0 - 0.169\,2\,R_{15}^{1/2} + 25.385\,1\,R_{15} + 14.094\,1\,R_{15}^{3/2} \\ - 7.026\,1\,R_{15}^2 + 2.708\,1\,R_{15}^{5/2} \tag{3.3}$$

1   You do not need to remember any details of equation 3.3. Question 3.7 (opposite) is intended simply to show you how it works.

2   Because the definition is a ratio, salinities should be presented simply as numbers, as in this text; but you may find salinities quoted in terms of *practical salinity units (p.s.u.)*. It is important to remember that the numbers approximate closely to grams per kilogram (or grams per litre), i.e. parts per thousand by weight.

3   In practice, computer algorithms are used for converting conductivity ratios at temperatures and pressures of measurement other than 15 °C and 1 atmosphere into R_{15}, and for the direct conversion of R_{15} into S.

QUESTION 3.7   Use equation 3.3 to answer this question by completing the following sentence. By definition, when $R_{15} = 1$, the practical salinity is exactly equal to … ?   *35*

A salinity value determined by conductivity depends on the temperature and pressure at which the conductivity is measured, and is thus somewhat removed from the simple but fundamental idea of salinity being the total dissolved salts in a seawater sample. In fact, for open ocean seawater, the two are closely related: the concentration of total dissolved salts in g per kg of seawater is $1.005\,10 \times S$, where S is as defined in equation 3.3.

3.4   SUMMARY OF CHAPTER 3

1   The average salinity of seawater is close to 35 parts per thousand (‰) by weight. Eleven major ions make up 99.9 per cent of the dissolved constituents: Cl^-, Na^+, SO_4^{2-}, Mg^{2+}, Ca^{2+}, K^+, HCO_3^-, Br^-, $H_2BO_3^-$, Sr^{2+} and F^-, in that order. The relative proportions of elements in solution in seawater differ greatly from the proportions in crustal rocks, because of their different solubilities in the solutions formed during terrestrial weathering and sea-floor hydrothermal activity.

2   Salinity varies from place to place in the oceans, but the relative proportions of most major dissolved constituents (their ionic ratios) remain virtually constant. Evaporation and precipitation change the total salinity, but do not affect the constancy of composition.

3   Minor departures from constancy of composition in the open oceans result mainly from the intervention of biological processes, affecting principally Ca^{2+}, and HCO_3^-. Major departures are the result of local conditions, chiefly in shallow nearshore waters and under anoxic conditions, but also where hydrothermal activity occurs. Some dissolved constituents are extracted commercially from seawater.

4   As in the case of temperature, the vertical and lateral distributions of salinity in the oceans do not change significantly from year to year, but the waters themselves are continually moving in a three-dimensional system of surface and deep currents. Surface salinities in the open oceans are greatest (up to 38) in tropical and subtropical latitudes, where evaporation exceeds precipitation. They are lower near the Equator (*c.* 35) and in high latitudes (*c.* 33–34), because of greater rainfall and melting ice and snowfall. In middle and low latitudes, there is a halocline from the base of the mixed surface layer to about 1 000 m depth, below which salinities are generally between 34.5 and 35.

5 Gravimetric measurement of salinity is difficult because of decomposition of some salts on heating to evaporation. Chemical measurements of salinity, based on titration to determine chlorinity, were standard until the 1960s, but have been almost entirely superseded by electrical conductivity methods. An empirically determined formula is used to convert conductivities, measured against a standard, into salinities.

Now try the following questions to consolidate your understanding of this Chapter.

QUESTION 3.8 Which of the following statements (a)–(e) are true, and which are false?

(a) The relative proportions of elements dissolved in seawater are very similar to those in average crustal rocks.

(b) Salinity can vary from place to place in the oceans, but the ratio of salinity to chlorinity will nearly always remain constant.

(c) The ratio of Ca^{2+} to salinity will fall where there is significant precipitation of calcium carbonate.

(d) Haloclines are regions in which salinity increases with depth.

(e) It is impossible to measure salinity and temperature to better than ± 0.01 and $0.01\,^{\circ}C$ respectively.

QUESTION 3.9 The oceans are not a closed system, and large amounts of dissolved salts are continually being introduced into them from the world's rivers. There are also significant inputs from hydrothermal solutions. So how is it that, in general, the constancy of composition of seawater is maintained?

CHAPTER 4

DENSITY AND PRESSURE IN THE OCEANS

The vertical and horizontal distributions of isotherms and isohalines generally remain fairly constant from year to year; seasonal fluctuations are largely confined to the surface layer. We have emphasized that these distributions represent a form of *dynamic equilibrium* or steady state, because the ocean waters themselves are continuously moving. The motion is not random, but is organized in a three-dimensional circulation system that shows little variation when motions are averaged out over periods of several years.

4.1 WATER MASSES

The Earth's climate and weather are largely the result of the movements of large air masses, each characterized by particular and recognizable combinations of temperature, humidity and pressure. In much the same way, large **water masses** in the ocean move vertically and horizontally, each defined by its temperature (T), salinity (S) and other characteristics, which can be used to identify it and track its movements. The main features of the movement of water masses are summarized below:

1 Figure 4.1 shows the boundaries of water masses formed in upper parts of the oceans, extending from surface or near-surface waters down to about the base of the permanent thermocline. They are identified by their temperature, salinity and other properties, including the communities of organisms that inhabit them. If you compare Figure 4.1 with Figure 2.11, you can see that the boundaries between these upper water masses coincide quite well with major surface current systems. It is also possible to identify boundaries between water masses moving in different directions at greater depths in the oceans.

QUESTION 4.1 (a) Look at Figure 2.13 and try to sketch in the boundaries between three water masses below the 10 °C isotherm.

(b) Water masses can be identified by their temperature (T) and salinity (S) signatures. In what way would you expect these properties to change (i) within and (ii) at the boundaries of water masses?

2 Water moves much more slowly than air, so water masses are less variable than air masses, and their boundaries do not change much, even on time-scales of decades to centuries.

3 The surface current systems are driven by winds, but the movement of intermediate and deep water masses is controlled mainly by density. When the density of the surface layers of seawater is sufficiently increased, the water column becomes gravitationally unstable and the denser waters sink.

QUESTION 4.2 (a) How can the density of surface waters in the oceans be increased in (i) polar and (ii) tropical regions?

(b) Is it reasonable to regard density-driven circulation in the ocean depths as a consequence of interactions between the atmosphere and oceans?

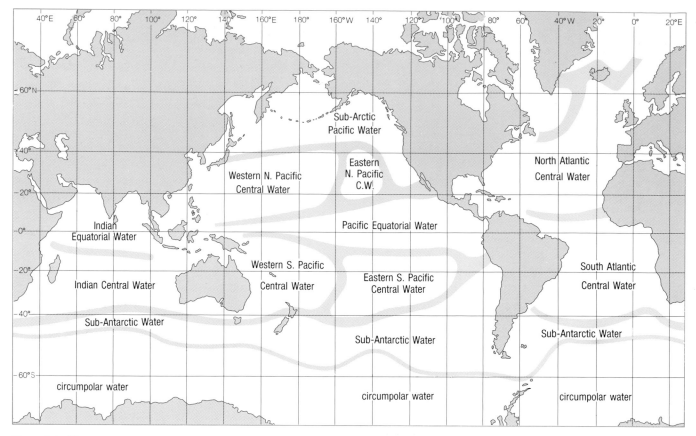

Figure 4.1   The approximate boundaries of the main upper water masses of the oceans (cf. Figure 2.11).

Vertical circulation in the oceans is controlled by variations in both temperature and salinity, and it is known as the **thermohaline circulation**. Its principal components are the cold dense water masses produced at high latitudes, which sink and spread throughout the oceans beneath the permanent thermocline, cf. Figure 2.13. Each of these water masses has a characteristic T and S signature inherited from surface conditions in its source region. Deep water from the Antarctic (Antarctic Bottom Water AABW, see Figure A1, with the answer to Question 4.1) crosses the Equator into the Northern Hemisphere. In the North Atlantic, there are comparable southward-flowing deep currents coming from the Arctic, but there are no such currents in the North Pacific, partly because of the barrier formed by the Aleutian island arc to the north.

In the surface waters of the oceans, temperature and salinity alone control the density of seawater, but in the deep oceans another factor becomes important: pressure.

4.2   DEPTH (PRESSURE), DENSITY AND TEMPERATURE

The density of seawater does vary somewhat with depth, but not to the extent supposed barely a century-and-a-half ago (Figure 4.2):

'The enormous pressure at these great depths seemed at first sight alone sufficient to put any idea of life out of the question. There was a curious popular notion, in which I well remember sharing when a boy, that, in going down, the seawater became gradually under the pressure heavier and heavier, and that all the loose things in the sea floated at

Figure 4.2
'The wrecks dissolve above us;
their dust drops down from afar –
Down to the dark, to the utter dark …'

Rudyard Kipling, *Song of the English.*

different levels, according to their specific weight: skeletons of men, anchors and shot and cannon, and last of all the broad gold pieces wrecked in the loss of many a galleon on the Spanish Main; the whole forming a kind of "false bottom" to the ocean, beneath which there lay all the depth of clear still water, which was heavier than molten gold.'

C. Wyville Thomson (1873) *The Depths of the Sea*, Macmillan.

The effect of pressure on density is not quite so dramatic as that. However, we should not scoff at such notions. Indeed, the concept of neutral buoyancy which is implicit in them is applied in modern technology (see Section 5.2.3).

The **hydrostatic equation** describes the way in which pressure P is related to depth (or height) (z) in a column of fluid:

$$P = g\rho z \tag{4.1}$$

where g is the acceleration due to gravity and ρ (rho) is the density.

Provided density remains constant, the hydrostatic equation shows a proportional relationship between pressure and depth (height). It is generally valid for the oceans, because water is only slightly compressible and the density of 99% of seawater is within $\pm 2\%$ of its mean value of about $1.03 \times 10^3 \, \mathrm{kg\,m^{-3}}$. On the scale of Figure 4.3, the result is a straight line.

QUESTION 4.3 (a) The value of g is $9.8 \, \mathrm{m\,s^{-2}}$. Use the hydrostatic equation to work out the pressure due to a 10 m column of seawater. Your answer will be in $\mathrm{N\,m^{-2}}$ (newtons per square metre). How does your answer compare with the value for normal atmospheric pressure?

(b) Approximately what pressures characterize (i) most of the deep ocean floors and (ii) the ocean trenches?

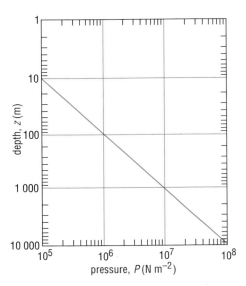

Figure 4.3 Graph of pressure (P) against depth (z) in the oceans. Both scales are logarithmic simply to accommodate the range of numbers. The relationship between pressure and depth is effectively linear when plotted on this scale. (Pressure is measured in newtons per square metre; $10^5 \, \mathrm{N\,m^{-2}} = 1 \, \mathrm{bar} \approx 1$ atmosphere.)

4.2.1 ADIABATIC TEMPERATURE CHANGES

As you read in Section 1.2.1, *adiabatic changes* of temperature are those that occur independently of any transfer of heat to or from the surroundings. They are a consequence of the compressibility of fluids. When a fluid expands, it loses internal energy and its temperature falls. When compressed, it gains internal energy and its temperature rises – that is the principal reason why the pump heats up when you inflate your bicycle tyres. The principles of adiabatic gain and loss of heat on compression and expansion of gases provide the basis of refrigeration and air conditioning technology. As air rises into a region of lower pressure it expands, and the rate of fall of temperature for air is c. 8–10 °C km^{-1}, depending on humidity (moisture content). Liquids are much less compressible than gases, and the rate of change of temperature with depth in the oceans as a result of adiabatic changes is less than 0.2 °C km^{-1}.

This brings us to the very important concept of **potential temperature**, θ (theta). In both the oceans and the atmosphere it is defined as the temperature which the fluid would attain if brought adiabatically from the pressure appropriate to its actual height or depth to a pressure of 1 000 millibars (i.e. approximately one atmosphere at sea-level). It is thus different from the ***in situ* temperature**, which is the temperature of the fluid measured at its actual height or depth.

42

Because of the great contrast in compressibility, the difference between potential and *in situ* temperature may be tens of degrees in the atmosphere, but is never more than about 1.5 °C in the oceans. The latter figure may seem to be trivial, but you will soon see that potential temperature is a very important concept when we come to consider vertical temperature distribution and gravitational stability in the oceans.

QUESTION 4.4   Explain whether you would expect the potential temperature of (a) air at a height of 5 km and (b) seawater at a depth of 5 km to be greater or less than their respective *in situ* temperatures.

4.3   *T–S* DIAGRAMS

***T–S* diagrams** are used to plot *in situ* temperature and salinity data for water samples, and hence to identify water masses. Figure 4.4 is a *T–S* diagram. The contours are lines of equal density. The numbers are values of σ_t (**sigma-t**), which is a convenient shorthand widely used in physical oceanography.

4.3.1   USING σ_t

σ_t is a shorthand way of expressing the density of a sample of seawater at *atmospheric pressure*, as determined from its temperature measured *in situ*

Figure 4.4   *T–S* diagram with contours of σ_t, which has units of density, kg m⁻³.

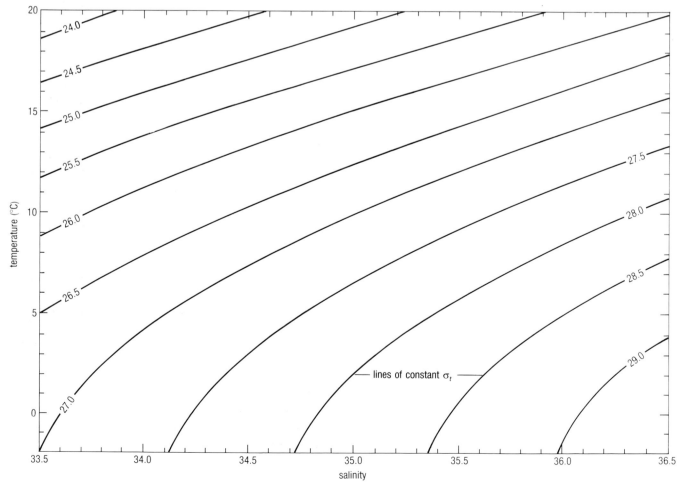

and its salinity. For example, on Figure 4.4 the σ_t of seawater at 5 °C (*in situ* temperature) and salinity 33.5 is 26.5 kg m^{-3}. The density of that water is $1.026\ 5 \times 10^3$ kg m^{-3}.

In general terms:

$$\sigma_t = (\rho - 1\ 000)\ \text{kg m}^{-3} \tag{4.2}$$

and so σ_t is sometimes known as the density anomaly.

The definition in equation 4.2 is relatively new. You may find σ_t values elsewhere quoted without units, because that was conventional practice until the late 1980s.

QUESTION 4.5 To make sure you understand the use of σ_t, attempt these simple questions.

(a) What are the values of σ_t for seawater at (i) *in situ* temperature of 2 °C and salinity of 34.5, and (ii) *in situ* temperature of 15 °C and salinity of 35.6?

(b) What do those σ_t values mean in terms of density at atmospheric pressure?

Taking the oceans as a whole, the range of temperatures is of the order of 0–25 °C (Figure 2.1), whereas that of salinity is generally little more than 34–36 (Figures 3.3 and 3.4), and can be less in individual ocean basins (see Figure 4.18). Temperature therefore typically influences density more than salinity does, e.g. for temperatures greater than about 5 °C a temperature change of 1 °C affects density more than does a salinity change of 0.1.

Where might you expect to find exceptions to this generalization?

In equatorial and high latitudes, where seasonal temperature changes are not so great (Figures 2.3(c) and 2.12), evaporation/precipitation and ice formation/melting can cause significant variations of salinity – and hence of density – in surface waters.

We have seen that below depths of *c.* 500–1 000 m in the oceans, temperature and salinity do not vary much. Figure 4.5(a) (overleaf) shows how this is reflected in the rather small increase of σ_t with depth below about 1 000 m. The σ_t profile is almost vertical below about 2 000 m.

In contrast, at depths of less than 500 m in middle and low latitudes σ_t increases rapidly with depth below the mixed surface layer, and the curves in Figure 4.5(a) are almost horizontal. A marked step in a density profile is termed a **pycnocline**. In the open oceans, pycnoclines are usually associated with thermoclines, though their exact position and slope will also depend upon the distribution of salinity. The main pycnocline coincides approximately with the permanent thermocline. Water in a pycnocline is necessarily very stable, i.e. it takes a large amount of energy to displace it up or down. The main pycnocline forms a lower limit or 'floor' to turbulence caused by mixing processes at the surface. Indeed, the very process of mixing in the surface layer tends to increase stability at its base, with development of a pycnocline (Figure 4.5(b)).

The depth of the mixed surface layer (Section 2.3) depends on the strength of the wind and on the processes that tend to promote vertical gravitational stability, such as heating of the surface and precipitation.

How do heating and precipitation promote stability?

Both reduce the density of surface waters: warm water is less dense than cold water and freshwater is less dense than seawater.

A gravitationally stable water column is said to be *stratified*, consisting of layers (strata) of water whose density increases with depth – boundaries between layers are typically gradational, but can be sharp (see Section 4.4.2). There are of course degrees of **stratification**, and hence of stability: a strongly stratified water column (rapid increase of density with depth) is more stable than a weakly stratified one (gradual increase of density with depth). A well-mixed water column (e.g. the mixed surface layer) is by definition not stratified and even small perturbations (e.g. turbulence, advection into it of water of different T or S) can easily make it unstable and lead to vertical mixing.

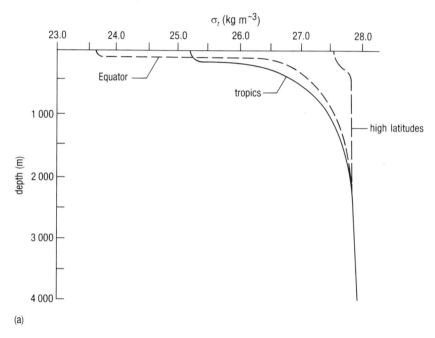

(a)

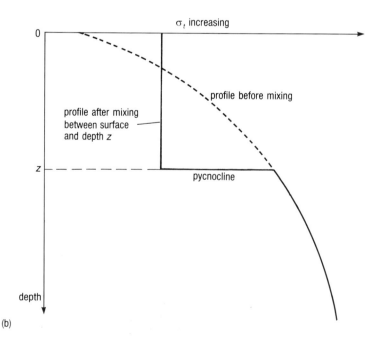

Figure 4.5 (a) Profiles of σ_t for different latitudes. Regions in which density changes sharply with depth are known as pycnoclines. Coincidence of the three curves below about 2 000 m results from regional variations of temperature and salinity in the deep oceans being relatively small.

(b) Formation of the mixed surface layer changes the density profile, with development of a pycnocline (see (a)) at the base of the mixed layer. Note the enlarged scale compared to (a).

(b)

4.3.2 σ_θ AND VERTICAL STABILITY

T–S diagrams are extremely useful for identifying and tracing water masses in the oceans, but they can give a spurious impression of gravitational instability in deep waters.

Bearing in mind what you read earlier about potential temperature and how σ_t is determined, can you see how this could happen?

Density must increase with depth, to ensure gravitational stability in the oceans. Adiabatic compression raises the temperature of deep water, so that *in situ* temperature becomes progressively greater than potential temperature with increasing depth. But σ_t is determined using *in situ* temperature uncorrected for adiabatic changes, so it will represent a density lower than that actually possessed by the water at any particular depth. In some cases, the differences are small enough to ignore, but it can happen that plots of salinity and *in situ* temperature show a decrease of σ_t with increasing depth, especially for deep water samples. These apparent instabilities disappear when the potential temperature, θ (Section 4.2.1) is used with salinity on a θ–S diagram to determine values of σ_θ (**sigma-theta**), and hence of **potential density** (i.e. the potential density anomaly $\sigma_\theta = $ (potential density − 1 000) kg m^{-3}).

Recalling that potential temperature is defined as the temperature which a sample of water would have if brought adiabatically from depth to atmospheric pressure, would you expect σ_t and σ_θ for surface-water samples to be the same?

As water at the surface is under atmospheric pressure, there is no need to make the adiabatic correction, so σ_t and σ_θ for surface-water samples must be the same.

Table 4.1 shows how, in the Mindanao Trench off the Philippines, σ_t (calculated from observed salinity and *in situ* temperature measurements) increases down to 4 450 m and then decreases again. This appears to suggest that the water column is gravitationally unstable. However, when *in situ* temperatures are converted to potential temperatures, σ_t is replaced by σ_θ and the apparent instability disappears.

Table 4.1 Comparison of *in situ* and potential temperatures in the Mindanao Trench off the Philippine Islands. See also Figure 4.6.

Depth (m)	Salinity	Temperature		Density	
		in situ (°C)	potential (°C)	σ_t (kg m^{-3})	σ_θ (kg m^{-3})
1 455	34.58	3.20	3.09	27.55	27.56
2 470	34.64	1.82	1.65	27.72	27.73
3 470	34.67	1.59	1.31	27.76	27.78
4 450	34.67	1.65	1.25	27.76	27.78
6 450	34.67	1.93	1.25	27.74	27.79
8 450	34.69	2.23	1.22	27.72	27.79
10 035	34.67	2.48	1.16	27.69	27.79

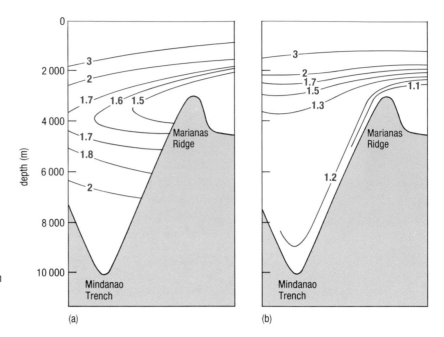

Figure 4.6   Two patterns of temperature distribution in the Mindanao Trench (for use with Question 4.6). Contours are in °C and represent either *in situ* temperature, or potential temperature, θ. See also Table 4.1.

(a)   (b)

QUESTION 4.6   'Cool water flows over the sill formed by the Marianas Ridge (Figure 4.6) and down the slope to the bottom of the Mindanao Trench.' From this description, which of the two diagrams in Figure 4.6 must show the potential temperature contours?

Table 4.1 shows that the difference between *in situ* and potential temperature exceeds 1 °C below depths of 8 km, while even at depths of about 1 km it is nearly a tenth of a degree. The differences clearly become much smaller as depth decreases, but it is important to recognize that there are small adiabatic temperature gradients even in a mixed surface layer which is otherwise isothermal. The differences may be small, but the sensitivity of modern instruments means that in some circumstances it is worth making the correction for adiabatic effects even in the top 200 m of the oceans. Moreover, modern technology enables potential temperature, θ, to be obtained automatically from *in situ* temperature measurements, and σ_θ is increasingly used in preference to σ_t.

4.3.3   THE USE OF *T–S* DIAGRAMS

In Section 4.1, you read that water masses can be identified by their *T–S* signatures. For example, near their source regions the three major subsurface water masses in the Atlantic Ocean (whose boundaries are approximately delineated in Figure A1, with the answer to Question 4.1) are characterized by the following narrow ranges of temperature and salinity:

Antarctic Bottom Water (AABW)	−0.5 ° to 0 °C and 34.6 to 34.7
North Atlantic Deep Water (NADW)	2 ° to 4 °C and 34.9 to 35.0
Antarctic Intermediate Water (AAIW)	3 ° to 4 °C and 34.2 to 34.3

T–S diagrams can thus be used both to identify water masses and to determine the extent to which they are mixed with one another. For example, Figure 4.7 is a *T–S* diagram on which *T* and *S* data for a station in the southern equatorial Atlantic have been plotted. The *T* and *S* 'signatures' of the three water masses quoted above are also shown.

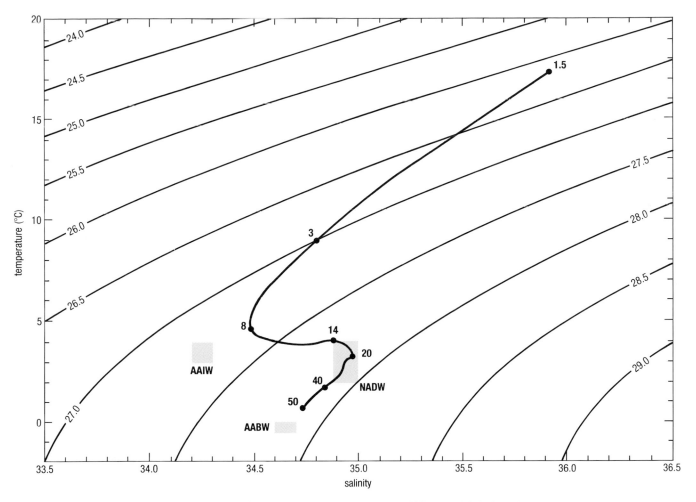

Figure 4.7 An example of a *T–S* diagram for observations from 150 m to 5 000 m depth at a location 9° S in the Atlantic Ocean. Contours are σ_t (kg m^{-3}). Dots represent individual seawater samples; numbers are depths in hundreds of metres. Blue shaded boxes represent the major subsurface Atlantic water masses.
AABW = Antarctic Bottom Water;
NADW = North Atlantic Deep Water;
AAIW = Antarctic Intermediate Water.

The water between about 1 400 m and 3 800 m depth represents NADW, scarcely modified at all by mixing, even at the low latitude of this station (9° S). For simplicity, we are treating NADW as a single water mass, but in fact it comprises more than one, with source regions mainly in the Norwegian and Greenland Seas.

The influence of AABW is identifiable at the bottom of the *T–S* curve in Figure 4.7, even though this bottom water has travelled thousands of kilometres from its source region in Antarctica. By contrast, the water at around 800 m depth still shows some of the features of AAIW, but this water mass has been considerably 'degraded' by mixing with surface water above and with deeper water below.

QUESTION 4.7 (a) If you were to plot the σ_t values on Figure 4.7 against depth, would the result indicate that the water column is gravitationally stable?

(b) Why would this result be only a rather rough indication of stability? Would a plot of σ_θ against depth give a more reliable indication?

(c) How would the curve on Figure 4.7 look if potential temperature were used to plot the points instead of *in situ* temperature? (Assume that in this case the contours would be for σ_θ values.)

Following on from Question 4.7, always remember that increasing density on a *T–S* curve corresponds to increasing depth. Hence, on diagrams such as Figure 4.7, where a *T–S* curve crosses contours in such a way that density increases with depth, the water column must be gravitationally stable (cf. the discussion of pycnoclines, Figure 4.5 and related text). Moreover, the greater the rate of increase of density with depth, the stronger the stratification, and the more gravitationally stable the water column will be: i.e. water is most stable within a pycnocline. By contrast, where the $T–S$ curve is roughly parallel with the contours, the density must be virtually uniform throughout that part of the water column, i.e. the water is well mixed (unstratified) and therefore not very stable.

Finally in this Section, we should note that as seawater is slightly compressible (Section 4.2), the *true density* of a seawater sample must be even greater than its potential density, because both σ_t and σ_θ are normally determined assuming atmospheric pressure (though potential density can be defined with reference to any selected pressure, say 200 atmospheres, equivalent to a depth of about 2 000 m). The compressibility of seawater also means that the true density increases slightly with depth (it is about 4% greater at 10 000 m depth than at the surface). So, according to the hydrostatic equation (4.1), on a larger and more detailed version of Figure 4.3, the graph would diverge slightly but progressively from the straight line with increasing depth, giving it a gently concave form.

σ_t, σ_θ, *and* γ

The equation used to determine the density of seawater from temperature, salinity and pressure has been refined in recent years. As a result, slightly different values are obtained for density (ρ) and hence for density anomaly ($\rho - 1\,000$). In the early 1980s, it was proposed that the symbol σ (sigma) be replaced by γ (gamma) to reflect these differences. This change is only very slowly gaining acceptance within the oceanographic community, which is why we have also continued to use σ rather than introduce γ – and anyway, numerical differences between σ and γ values are very small.

4.3.4 CONSERVATIVE AND NON-CONSERVATIVE PROPERTIES

There are two main reasons why the $T–S$ diagram is a powerful tool for identifying and tracking water masses. First, temperature and salinity are quite easily measured. Secondly, as soon as the water is out of contact with the atmosphere, i.e. it has left the mixed surface layer and is in the main body of the ocean, *these properties can only be changed by mixing with water of different T and S characteristics.* For this reason, T and S are known as **conservative properties.**

Bearing in mind the definition you have just read, would you say that, strictly speaking, potential temperature, θ, is a true conservative property, whereas *in situ* temperature is not?

Strictly speaking, the answer is yes. *In situ* temperature can be changed by processes other than mixing, namely by adiabatic compression or expansion. Potential temperature has been corrected for this effect, so it is the true conservative property. $T–S$ diagrams are being increasingly replaced by $\theta–S$ diagrams, which are used in *exactly* the same way as outlined in Section 4.3.3 for Figure 4.7.

Water masses can be identified also by chemical and biological characteristics, for example by their content of dissolved oxygen or nutrients; and, in the case of upper water masses in particular (cf. Figure 4.1), by the presence of certain communities of organisms. Clearly, however, all of these properties can be changed by processes other than mixing, especially biological processes, and so they are called **non-conservative properties**.

It is crucial to remember that these definitions only apply away from boundaries with the atmosphere and sea-bed. At these boundaries, there are gains or losses of heat, salt or freshwater, by solar radiation, rainfall, river inflow, crustal heat flux, and so on. The distinction between conservative and non-conservative properties and behaviour is extremely important in oceanography and you will encounter applications again in Chapter 6.

Hydrothermal circulation associated with sea-floor spreading and submarine volcanism supplies large volumes of heated water into oceanic bottom waters, especially at spreading axes. Does this process invalidate the definition of conservative properties given above?

Not at all. The water expelled from hydrothermal vents has very different values of temperature and salinity from those of surrounding bottom waters. Temperature and salinity are conservative properties, and so they can be used to track the subsequent movements of hydrothermal waters in just the same way as is done for the major water masses.

QUESTION 4.8 (a) Look back to Section 3.1 and recall the discussion of departures from constancy of composition for some major constituents. Do those constituents behave conservatively or non-conservatively in seawater?

(b) Is chlorinity a conservative property?

In fact, amounts of Ca^{2+} removed from solution by biological processes are small in relation to its total concentration (Section 3.1) and calcium is generally treated as conservative by most oceanographers (see Sections 6.1.1 and 6.1.2). Changes in the concentration of bicarbonate (HCO_3^-) are greater, however (Section 3.1), so this constituent is generally classed as non-conservative.

4.4 MIXING PROCESSES IN THE OCEANS

Inhomogeneities in the oceans can occur on a variety of scales, the largest being that of water masses mentioned earlier in this Chapter. We look at smaller scale inhomogeneities later in this Section. Mixing processes act to even out the inhomogeneities: they encompass the extremely slow process of molecular diffusion and the much more rapid process of turbulent mixing.

4.4.1 MOLECULAR AND TURBULENT DIFFUSION

Even in a fluid that is absolutely at rest, if a dissolved substance is unevenly distributed within it, the substance will diffuse down the concentration gradients to even out the distribution. This is **molecular diffusion**, resulting from the motion of individual molecules. An even distribution of heat is achieved in a similar way: in regions of higher temperature the molecules have higher kinetic energies. Molecular diffusion of heat occurs as these higher energy molecules move (diffuse) down temperature gradients into regions of lower temperature where they encounter slower-moving molecules and transmit some of their excess energy to them. This is how the process of conduction operates *in a fluid*.

Water in the oceans is usually moving, rarely in laminar flow, most commonly in turbulent fashion. The distinction between the two is shown in Figure 4.8.

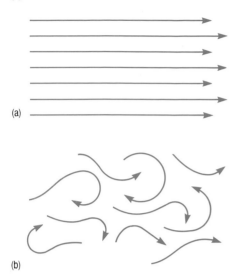

Figure 4.8 Diagrammatic illustration of the difference between (a) laminar flow and (b) turbulent flow.

When fluid is moving by laminar flow, mixing occurs mainly by molecular diffusion. Turbulence (Figure 4.8(b)) can bring waters with very different characteristics into close proximity. It involves bulk mixing, like sloshing the water about in a bath, which very quickly achieves a uniform temperature and an even distribution of bath salts. In the oceans, therefore, mixing occurs mainly by **turbulent diffusion**, which is many orders of magnitude faster than molecular diffusion. Irrespective of whether mixing is by molecular or turbulent diffusion, however, the diffusion must take place 'down the gradient' of temperature or concentration, i.e. from *higher* to *lower* temperature or from *higher* to *lower* concentrations of dissolved salts, nutrients, dissolved gases, and so on. As noted above, rates of turbulent diffusion are much greater than those of molecular diffusion.

In the oceans, turbulence may be associated with a wide range of processes: wind-driven wave motions; convective overturn caused by density differences; vertical or lateral **current shear** (i.e. variations of velocity either with depth or across the flow); water movement over an irregular sea-bed or along an irregular coast; tidal currents, which vary with time as well as with position; and travelling eddies associated with currents (see Section 4.4.4).

The oceans are much broader than they are deep – up to about 10 000 km across, compared with about 5 km deep – and horizontal gradients of temperature are several orders of magnitude less than the corresponding vertical gradients. Temperature can change by 10 °C or more in 1 km depth, whereas it is commonly necessary to travel thousands of kilometres horizontally to experience a temperature change of 10 °C. The scale of horizontal turbulent mixing is greater than that of vertical turbulent mixing, which tends to be opposed by the vertical gravitational stability that results from the increase of density with depth. In short, the effect of density stratification is to inhibit or suppress vertical mixing.

4.4.2 STRATIFICATION AND MICROSTRUCTURE

Instruments that can provide continuous profiles of temperature and salinity in the oceans reveal fine-scale stratification features that are known as **oceanic microstructure**. Step-like profiles, in which homogeneous layers of water are separated by thin interfaces with steep gradients of temperature and salinity (Figure 4.9), have been found in many regions. The scale of these features varies considerably, some layers being 20–30 m thick (Figure 4.9(a)), whereas others, perhaps superimposed on them, are only 0.2–0.3 m thick (Figure 4.9(c) and (d)). Their lateral extent may be as much as tens of kilometres for thicker layers and perhaps hundreds of metres for thinner layers. Temperatures may either increase or decrease with depth in these step-like profiles, but where the temperature increases with depth (a temperature inversion) salinity also increases with depth, otherwise the interfaces between the layers would not be stable. Where the temperature decreases with depth, salinity may either increase or decrease with depth.

Because the density increases across each step, the microstructure is vertically stable, and this tends to inhibit vertical mixing. Molecular diffusion alone would eliminate differences between adjacent layers of water, given sufficient time. However, the persistence of sharp boundaries between the layers in oceanic microstructure suggests that there is some process which acts to maintain the contrasts across them, counteracting the effects of molecular diffusion. Various hypotheses have been proposed to account for oceanic microstructure and for the processes that act to maintain it. It may well be that different processes dominate on different scales in different parts

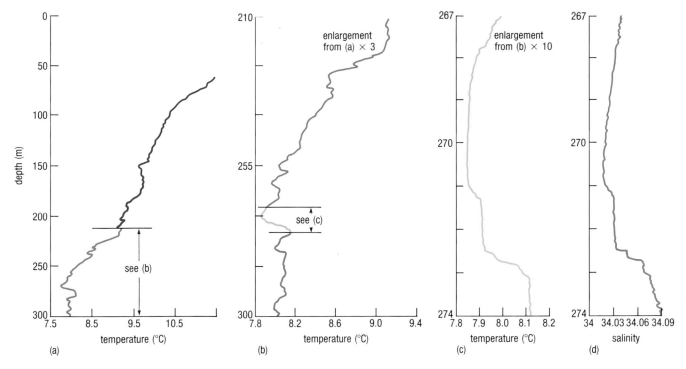

(a)   (b)   (c)   (d)

Figure 4.9   Step-like profiles of temperature – (a), (b), (c) – and salinity (d), from a location off the coast of California. Profiles (a)–(c) are successively expanded, to show the fine scale of stratification that can be detected. Microstructure can occur at any depth, but is most common within and above the main thermocline.

Figure 4.10   A gravitationally stable situation becomes unstable where warm saline water (light blue) overlies cooler and less saline water (dark blue), resulting in an additional step in the density profile. (a) The more rapid diffusion of heat (*short arrows*) than salt, leads to (b) and (c), the development of salt fingers (*long arrows*) when the density profile becomes unstable. (d) Schematic detail of a density profile, showing an extra step in a 'thermohaline staircase' formed after a salt fingering 'event'. Broken line before; solid line after.

of the oceans. Here, we describe two likely mechanisms for the maintenance of oceanic microstructure.

Salt fingering results from what is known as *double diffusion*, or double diffusive mixing of heat and salt. Molecular diffusion of heat is many times more rapid than that of salt. If, therefore, we initially have a two-layer system, where less dense warm salty water overlies denser, cooler and less salty water, the heat diffuses downwards more rapidly than the salt. Figure 4.10 shows how this process reduces the density of the lower layer and increases that of the upper layer, leading to instability in the system. The result is a convection pattern of sinking cells of salty water alternating with rising cells of less salty water.

The scale of these convection cells is only of the order of centimetres, so the effect of salt fingering is not to break down the stratification, but rather to create 'thermohaline staircases' of the kind illustrated in Figure 4.9 and to make the microstructure progressively finer and more detailed by adding intermediate steps (Figure 4.10(d)).

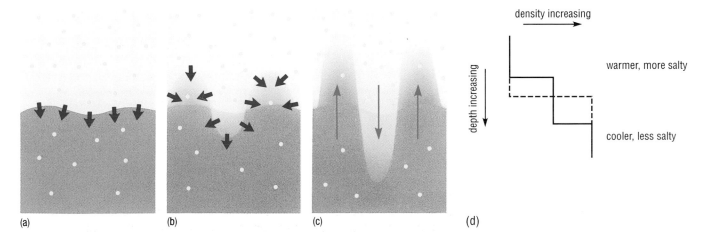

(a)   (b)   (c)   (d)

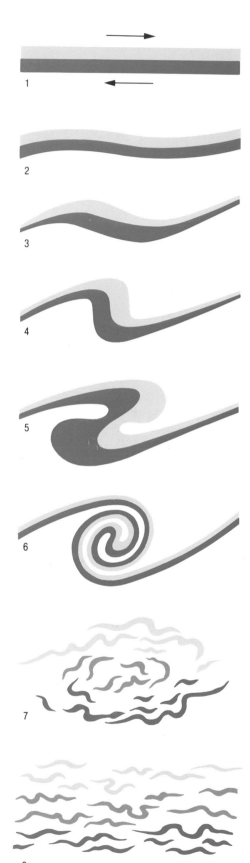

The breaking of internal waves

We have already established that the microstructure is gravitationally stable, as the density increases downwards across each step. Wherever the water is stable, oscillations can occur if it is displaced vertically. **Internal waves** result, which can propagate energy through the ocean in the same way as surface waves do.

Such waves can form at the interfaces between layers of different density which are associated with velocity shears, i.e. where the water above and below the interface is either moving in opposite directions or (more likely) in the same direction at different speeds. These shears can produce local instability in the form of billows or breakers (Figure 4.11) which lead to turbulent mixing of water immediately above and below the interface. As with salt fingering, the effect of this is to create an intermediate layer between the two original layers, and thus form two smaller steps in the vertical profile in place of one larger step. This can continue indefinitely with further steps in the vertical profile being formed on each occasion.

One of the first observations of this process was in the early 1970s, when the use of dye tracers enabled divers to observe internal waves breaking in the thermocline off Malta. Internal waves in general occur on a variety of scales and are a widespread phenomenon in the oceans. Probably the most important are those associated with tidal oscillations along continental margins. These are large enough to be detected easily on aerial photographs and satellite imagery, provided they are not too deep.

4.4.3 FRONTS

In the oceans, **fronts** are inclined boundaries between different bodies of water having contrasted characteristics. They are analogous to atmospheric fronts between different air masses, and occur on a variety of scales. They can form both within estuaries (between river water and higher salinity estuarine water) and off the mouths of estuaries (between estuarine water and normal seawater). They are common in shallow seas, separating stratified water from water that is vertically mixed; and along continental shelf margins, separating coastal or shelf water from water of the open ocean. On a still larger scale are fronts in the deep ocean between water masses of different properties, which often coincide with regions of strong current shear, as we saw when comparing Figures 2.11 and 4.1.

In shelf seas, tidal currents have appreciable velocities close to the sea-bed, and can be an important agent of vertical mixing. If there is a large vertical current shear due to friction at the sea-bed (Figure 4.12), the resulting vigorous turbulence leads to the development of a lower mixed layer. If the top of this lower layer merges with the base of the upper mixed layer, the water becomes vertically homogeneous – a common situation in the seas around Britain which are subject to fairly strong tidal currents ($>0.5\,\mathrm{m\,s^{-1}}$). In some areas, however, the tidal currents are

Figure 4.11 Mixing and microlayering caused by the passage of an internal wave. Stage 1 shows a layer of lower density overlying and moving faster than one of higher density, so that the *relative* speeds are in opposite directions (arrows). In the succeeding stages (2–8), the two layers lose their coherence as internal waves develop, and break into turbulent patches. The patches are rapidly flattened by stratification, which gives rise to a finely layered microstructure (cf. Figure 4.9) as the movement subsides.

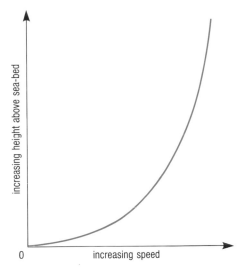

Figure 4.12 Variation in speed with height above the sea-bed to illustrate the principle of vertical current shear; each 'layer' moves faster than the one immediately below it.

weaker, or the total depth of water is greater, and in these areas stratification does develop in summer. Fronts in shelf seas are boundary regions between homogeneous (i.e. completely mixed) and stratified waters (Figure 4.13), where the balance between layering and mixing depends mainly upon the strength of the tidal currents.

Why is stratification more likely to develop in mid-latitude shelf-sea areas in summer than in winter?

Greater insolation in summer months leads to warmer and less dense surface waters, and mixing is not so intense because in general wind speeds are lower. A seasonal thermocline develops (Section 2.3). In winter, cold weather and generally stronger winds cool the surface layers, which become denser and less stable, and hence more susceptible to mixing by wind and waves. The thermocline is pushed deeper (cf. Figure 2.7(d)) and eventually intersects the top of the lower mixed layer; the whole water column will then be mixed throughout.

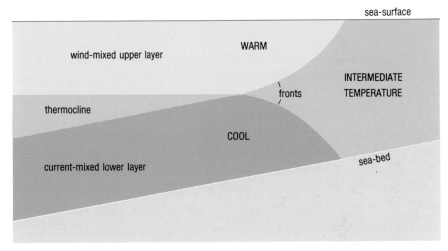

Figure 4.13   An illustration of how fronts can develop between homogeneous waters (right) and stratified waters (left) in shelf seas. (Note that the fronts will in general be gradational zones rather than sharply defined boundaries.) The lower mixed layer is due to tidal currents, the upper mixed layer is due to mixing by wind and its lower boundary is a seasonal thermocline (probably coinciding with a pycnocline). The two mixed layers merge and mix together where the water is shallower. Vertical scale greatly exaggerated.

The essential feature of a front is the density difference between the water on either side, but other features usually enable it to be seen fairly easily. The front itself is frequently marked by a line of foam or floating debris (Figure 4.14(a)), because fronts are regions where surface waters converge, i.e. move towards one another on either side of the boundary. The convergence results in part from wind at the surface, but it is also the result of density contrasts across the front.

Figure 4.14(b) illustrates in a simplified way how density contrasts may be associated with convergence and sinking of surface waters. Fronts by definition separate water of different density along inclined boundaries. There are strong density gradients *across* them, so fronts are defined by closely spaced (imaginary) surfaces of equal density – **isopycnal surfaces** (also known as isopycnic surfaces; *contours* of equal density are called **isopycnals**). Because isopycnal surfaces are inclined, the water 'slides'

(a)

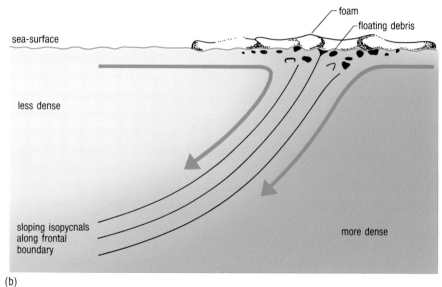

(b)

Figure 4.14   (a)   Typical foamline, the surface expression of a front (see (b)), at 42°20' N, 8°54' W.

(b)   Schematic illustration of the convergence and sinking of surface water along a frontal boundary. Note the strong density gradient *across* the front which is represented by sloping isopycnal surfaces. The vertical scale is greatly exaggerated. For further explanation, see text.

downwards along them. The sinking water draws more water in from above to maintain the supply. Sinking is generally more pronounced on the 'more dense' side of the front (Figure 4.14(b)), and denser water is sometimes said to be *subducted* beneath less dense water on the other side of the front.

Because the water properties on either side of a front are different, they can be identified easily on aerial photographs and satellite images, especially where there is a change in surface roughness and hence in optical reflectivity. The temperature of the water is nearly always significantly different on each side, and the less-stratified (better-mixed) cooler water on one side is more likely to be richer in nutrients than the more-stratified warmer water on the other. As a result, fronts can often be recognized by differences of biological production as well as of temperature, and the two are often well correlated (Figure 4.15).

As mentioned at the start of this Section, fronts can also develop in association with lateral velocity shears across current systems where

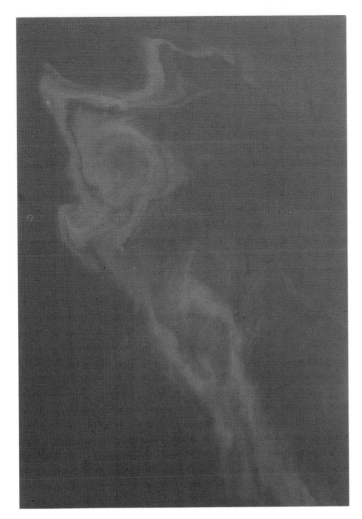

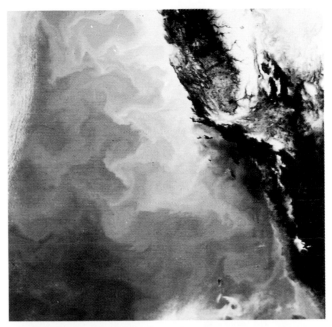

Figure 4.15 (a) A front in the Falklands Current, shown up by contrasted production of plankton. To the right of the front, the paler colours result from high populations of plankton (a phytoplankton bloom), while on the left, the ocean is relatively barren. The front shows the characteristic swirls and eddies found in nearly all frontal regions and at least in part related to the changes of current velocity (lateral velocity shears) across the front (see Section 4.4.4). The distance from top to bottom of the photograph is about 100 km.

(b) The correlation between sea-surface temperature (*top*) and biological productivity (phytoplankton, *bottom*), on each side of a front off southern California, near the edge of the continental shelf. The waters of the continental shelf are cooler and more nutrient-rich (because of mixing and run-off from land) than the stratified water offshore. This picture also shows how wave-like patterns and eddies can develop along fronts (see Section 4.4.4). The distance from top to bottom of each picture is 700 km.

adjacent bodies of water are moving in the same direction but at different speeds. They can be very sharp boundaries. For example:

> 'The Gulf Stream, as it issues from the Straits of Florida and expands into the ocean on its northward course, is probably the most glorious natural phenomenon on the face of the Earth. The water is of a clear crystalline transparency and an intense blue, and long after it has passed into the open sea it keeps itself apart, easily distinguished by its warmth, its colour, and its clearness; and with its edges so sharply defined that a ship may have her stem in the clear blue stream while her stern is still in the common water of the ocean.'

C. Wyville Thomson (1873) *The Depths of the Sea*, Macmillan, p. 382.

The boundaries of the Gulf Stream are normally not so abrupt as that: the 'cold wall', separating the warm waters of the Gulf Stream from cooler waters on the landward side, is typically a frontal *zone*, some 30–50 km wide, over which the temperature changes by as much as 10 °C.

Across most major fronts, however, the temperature gradient is typically much less: of the order of 2 °C in about 20 km. Smaller fronts in estuarine and coastal waters are sharper (cf. Figure 4.14). It is important to stress that the vertical scales in Figures 4.13 and 4.14 are highly exaggerated, because fronts typically slope at very low angles from the horizontal – of the order of 1 in 100.

As you might expect, mixing occurs across fronts, which is an important consideration in, for example, the exchange of coastal and open ocean waters, because the extent of mixing will control (among other things) the removal of pollutants to the deep ocean. Mechanisms of mixing include interleaving of the two water masses on either side of the front, producing a 'frontal microstructure' where small-scale mixing processes of the kind described in Section 4.4.2 might operate; and the development of eddies where there are velocity shears.

4.4.4 EDDIES

The swirls and eddies associated with fronts and currents (e.g. Figure 4.15) can occur on any scale and result from current shear across the flow. You can watch the development of small eddies for yourself in any reasonably fast-flowing river or coastal (tidal) current.

The **mesoscale eddies** which develop along major current systems, such as the Gulf Stream, bear the same relation to oceanic water masses that atmospheric depressions and anticyclones have to air masses, but are about ten times smaller (Figure 4.16). They typically have length and depth scales of the order of 100 km and hundreds to thousands of metres respectively, and time-scales ('lifetimes') of up to two years. Their existence was not even suspected until the 1960s and not confirmed till the 1970s, chiefly because they are difficult to detect and track using conventional shipboard techniques alone. Nowadays, they can be easily observed and monitored by satellite (Figure 4.17).

Mesoscale eddies play a major role in large-scale oceanic mixing processes, because they transfer substantial volumes of water of contrasted T and S, and other properties, from one side of a current system to the other (Figure 4.17).

Up to now, we have been concerned mainly with what may be called intrinsic properties of seawater. In the next Chapter we look at how these properties affect the propagation of light and sound in the ocean.

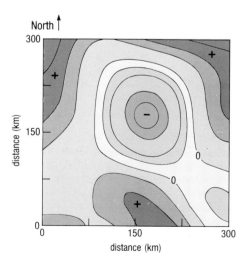

Figure 4.16   A mesoscale eddy 'mapped' by contours of temperature *difference* at 700 m depth in the north-western Atlantic. The zero contour represents the reference temperature, with blue and red colours representing respectively colder and warmer water. This type of eddy is often called a cold-core eddy.

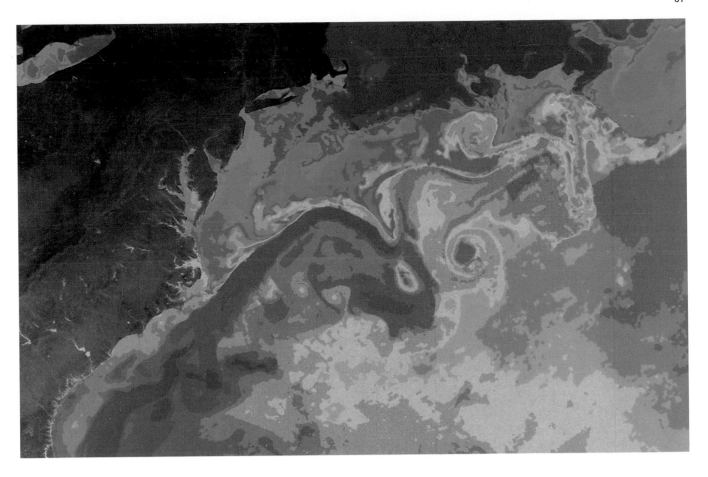

Figure 4.17 Infrared satellite image of mesoscale eddies 'spun off' from the Gulf Stream east of Cape Hatteras. Brown is land, red and purple respectively represent warmest and coldest surface waters; yellow, green and blue show intermediate temperatures. Two cold-core eddies have formed south of the Gulf Stream (green surrounded by yellow), and two warm-core eddies are in the process of forming north of it (yellow surrounded by green). These eddies have respectively transported cold water from the northern to the southern side of the Gulf Stream, and warm water from the southern to the northern side.

4.5 SUMMARY OF CHAPTER 4

1 Water masses are analogous to air masses. They can be identified by characteristic combinations of temperature and salinity and other properties. The boundaries of major upper water masses correspond approximately to the major wind-driven surface current systems. Subsurface water masses have comparatively narrow ranges of temperature and salinity, inherited from surface conditions in the source regions where they form and sink by virtue of their increased density. The movement of subsurface water masses is density-driven; this is the thermohaline circulation.

2 Temperature and salinity together control density, but pressure is also an important factor. Pressure increases almost linearly with depth in the oceans, because water is virtually incompressible. A pressure of about 1 atmosphere ($10^5\,\mathrm{N\,m^{-2}}$ or $1\,000$ mbar) is exerted by a $10\,\mathrm{m}$ water column. Air cools adiabatically as it rises, due to expansion as pressure falls. Water is heated adiabatically as a result of increased pressure and slight compression with depth. The potential temperature (θ) of a water sample is its measured *in situ* temperature after correction for adiabatic heating.

3 Sigma-t (σ_t) represents the density of seawater samples at atmospheric pressure, based on salinity and *in situ* temperature. Sigma-θ (σ_θ) represents the density of seawater samples at atmospheric pressure, based on salinity and potential temperature θ. T–S diagrams are contoured in values of σ_t and are used to identify water masses and to determine the extent of mixing between them. θ–S diagrams are contoured in values of σ_θ and are used in

exactly the same way. Pycnoclines are regions where density increases rapidly with depth, and the main pycnocline coincides approximately with the permanent thermocline.

4   Conservative properties of seawater are those that are changed only by mixing, once the water has been removed from contact with the atmosphere and other external influences. Non-conservative properties are those that are changed by processes other than mixing. Temperature (potential temperature) and salinity are conservative properties; dissolved oxygen and nutrient concentrations are non-conservative properties.

5   Mixing occurs by both molecular diffusion and turbulent diffusion, the second of which is by far the more important: turbulent diffusion is much more rapid than molecular diffusion. The scale of horizontal mixing is greater than that of vertical mixing in the oceans, partly because of their great width : depth ratio, and partly because density stratification inhibits vertical mixing.

6   In many parts of the ocean, there is a well-defined and gravitationally stable microstructure, consisting of layers of water with fairly uniform T and S characteristics, separated by steep gradients of temperature and salinity. Small-scale processes that may operate to form and maintain the stratification are salt fingering, which results from double diffusion of heat and salt; and breaking of internal waves, the result of velocity shears along density interfaces.

7   Fronts are gently inclined boundaries which separate water of contrasted characteristics, typically well stratified on one side, mixed and hence more uniform on the other. They are common in shallow continental shelf waters, over the continental shelf and along continental margins; and are associated with oceanic current systems. They are characterized by sloping isopycnal surfaces (surfaces of constant density). Near-surface water can sink to greater depths along sloping isopycnals. Major fronts are normally tens of km across and slope down beneath the warmer and more stratified water, often at very small angles.

8   Eddies can develop wherever there is velocity shear, and are commonly associated with fronts and currents. Mesoscale eddies which form along major current systems (e.g. the Gulf Stream) are an important agent of large-scale mixing in the oceans.

Now try the following questions to consolidate your understanding of this Chapter.

QUESTION 4.9   How do processes involved in the formation of sea-ice help to drive the thermohaline circulation of the oceans?

QUESTION 4.10   Temperature generally decreases with depth in the oceans. In the troposphere (the lower 10–15 km of the atmosphere), which contains three-quarters of the mass of the atmosphere, temperature decreases with height. Are the causes of the temperature decrease the same?

QUESTION 4.11   From Figure 4.4, for water at 16 °C and with a salinity of 34:

(a)   What change in *temperature* would cause a change in density from $1.025 \times 10^3 \, kg \, m^{-3}$ to $1.026 \times 10^3 \, kg \, m^{-3}$ at constant salinity?

(b)   What change in *salinity* would cause the same change at constant temperature?

(c) If a sample of seawater with T–S characteristics of 2 °C and 34.5 were collected from a depth of 4 000 m, would its σ_t value correspond to a density greater or less than its potential density?

QUESTION 4.12 Examine Figure 4.18. The large areas representing T and S values for the three main oceans (Pacific on the left, Atlantic on the right, Indian in the middle) converge near the bottom of the diagram.

(a) Does the large spread of values of T and S for each ocean represent mainly upper water masses or mainly deep water masses?

(b) Table 4.2 presents estimates of *average* temperature and salinity values for each of the main ocean basins, and the average for the world's ocean as a whole. Plot on Figure 4.18 the values for each ocean. Can you suggest why the three points cluster together where they do?

Table 4.2 Average temperatures and salinities for the major ocean basins.

Ocean	Temperature (°C)	Salinity
Pacific	3.36	34.62
Atlantic	3.73	34.90
Indian	3.72	34.76
All oceans	3.52	34.72

Figure 4.18 A *T–S* diagram for waters of the world's major ocean basins, excluding the mixed surface layer. (For use with Question 4.12.)

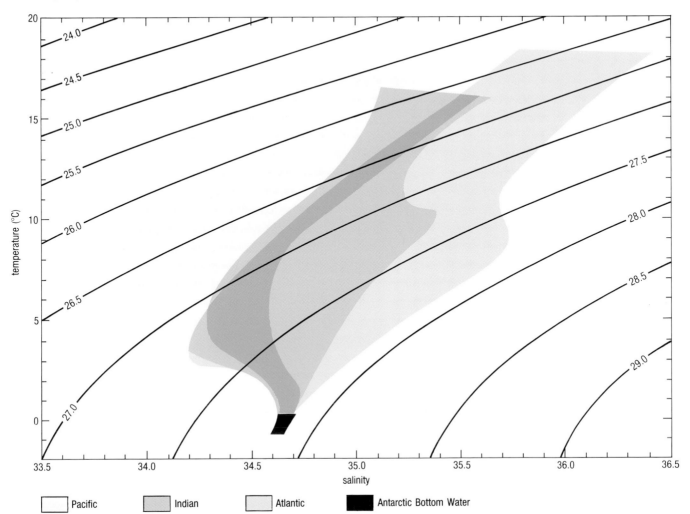

QUESTION 4.13   Are each of statements (a)–(e) true or false?

(a)   According to the definition, *in situ* temperature is not a conservative property, but potential temperature is.

(b)   When it is not possible to measure the *in situ* temperature, the potential temperature is measured instead.

(c)   Both surface heating of the oceans and precipitation (rainfall) promote stability of the surface layers.

(d)   A weakly stratified water column will be characterized by a steep thermocline and/or a steep halocline.

(e)   If the water column represented by the temperature profile in Figure 2.7(c) is gravitationally stable, surface water must be of lower salinity than water at 250 m depth.

QUESTION 4.14   This question concerns the formation of sea-ice and draws on material from both Chapter 1 and Chapter 4. As surface seawater cools, its density increases and it sinks (cf. Figure 1.2). In seawater of normal salinity, sea-ice can therefore only begin to form when the whole water column is cooled to about $-1.9\,°C$. This is not likely to happen in ocean basins that are several km deep. At first sight, then, it would seem that sea-ice could only form in shallow water; but in the Arctic and Antarctic Oceans it can form offshore where low salinity water overlies water of higher salinity.

(a)   By extrapolating from Figure 4.4, make an estimate of the density of water at a temperature of $-1\,°C$ and with a salinity of (i) 33 and (ii) 35.

(b)   Could the water of salinity 33 cool to freezing point without sinking below the higher salinity water beneath it?

CHAPTER 5   LIGHT AND SOUND IN SEAWATER

Humans are generally accustomed to consider sight to be a more important sense than hearing. Light travels faster and penetrates further through the atmosphere than does sound, so we can make better use of our sight, and of electromagnetic radiation generally, in making scientific observations. For animals in the oceans, by contrast, hearing is the more important sense. Sound travels well through water, and this makes possible the remote sensing of objects (e.g. echo-sounding) and the transmission of information (e.g. the 'singing' of whales). Light travels only relatively short distances through water, and the greater part of the oceans is almost completely dark.

5.1   UNDERWATER LIGHT

Light is a form of electromagnetic radiation, which travels at a speed close to 3×10^8 m s^{-1} in a vacuum (reduced to about 2.2×10^8 m s^{-1} in seawater). Oceanographers are interested in underwater light in two main contexts: vision and photosynthesis.

When light propagates through water, its intensity decreases exponentially with distance from the source; see Figure 5.1, *and note the very different depth and light intensity scales on (a) and (b)*. The exponential loss of intensity is called **attenuation** and it has two main causes:

1 **Absorption:**   This involves the conversion of electromagnetic energy into other forms, usually heat or chemical energy (e.g. photosynthesis). The absorbers in seawater are:

(a)   Algae (phytoplankton) using light as the energy source for photosynthesis.

(b)   Inorganic and organic particulate matter in suspension (other than algae).

(c)   Dissolved organic compounds (see Section 5.1.4).

(d)   Water itself.

Note: (a) and (b) are collectively termed the **seston** (cf. Section 6.1).

2 **Scattering:**   This simply changes the direction of the electromagnetic energy, as a result of multiple reflections from suspended particles. Scattering by all but the very smallest particles is generally forwards at low angles – i.e. the path of most of the scattered light is deflected only slightly from its original direction of propagation. Obviously, the greater the amount of suspended matter (i.e. the more turbid the water) the greater the degree of absorption and scattering.

Coastal waters tend to be particularly turbid. The suspended load brought in by rivers is kept in suspension by waves and tidal currents, which also stir up sediment already deposited on the bottom. In addition, rivers supply coastal waters with nutrients that support phytoplankton growth, and with dissolved organic compounds (item (c) in the list of absorbers above). By contrast, the water tends to be particularly clear in central oceanic regions, especially where concentrations of nutrients are low and where there is little biological production.

Figure 5.1 illustrates two ways in which the exponential decrease of illumination with depth can be represented. In (a), the upper picture, the horizontal scale for light intensity is *linear*. The curve shows how the exponential decrease of the intensity of sunlight with depth, even in clearest ocean water, is such that at about 250 m depth the light intensity has fallen by nearly three orders of magnitude, from 1 000 (10^3) W m^{-2} to little more than 1 W m^{-2}. In (b), the lower picture, light intensity is plotted on a *logarithmic* scale. The graph becomes linear and we can see the relationship

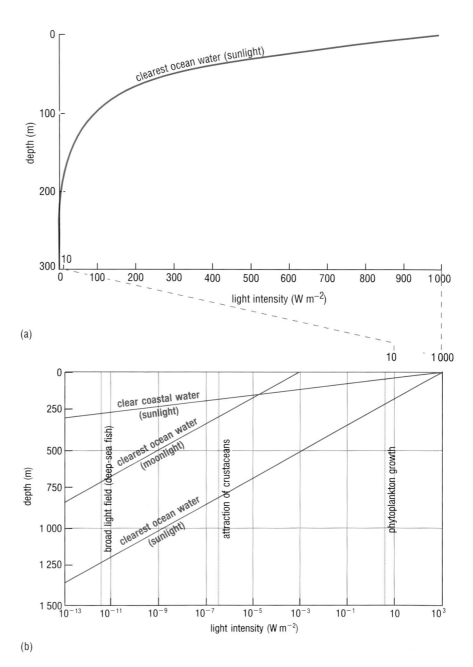

(a)

(b)

Figure 5.1   The relationship between illumination and depth in the ocean. (a) Light intensity plotted on a *linear* scale, down to 300 m depth. (b) Light intensity plotted on a *logarithmic* scale down to 1 500 m depth. The curve in (a) corresponds to the right-hand end of the *lowermost* diagonal line in (b). See text for further description.

between illumination and depth for light intensities less than $1\,W\,m^{-2}$, because the information between 10 and 0 on the horizontal scale in (a) is expanded into the space between 10 and 10^{-13} on the horizontal scale in (b). It is important to realize that the curved line in (a) has become the right-hand end of the *lowermost* diagonal line in (b), also that the depth scales on (a) and (b) are very different.

In the lower diagram (Figure 5.1(b)), pale blue vertical lines show the light intensity required for various functions. The 'broad light field' for deep-sea fish indicates the minimum quantity of general daylight that these fish can perceive. The intersection with the 'clearest ocean water (sunlight)' line at a point corresponding to about 1 250 m indicates that below this depth fish cannot perceive daylight. More light is needed to attract crustaceans, and still more for phytoplankton growth. (For comparison, the lowest intensities that the human eye can perceive are of the order of $10^{-12}\,W\,m^{-2}$ for a small light source and 10^{-8} to $10^{-9}\,W\,m^{-2}$ for a broad, diffuse light source.)

QUESTION 5.1 (a) According to Figure 5.1(b), is the intensity of light sufficient for phytoplankton to grow (i) on a moonlit night; (ii) at depths greater than 100 m in sunlit coastal waters; (iii) at depths less than 200 m in clear sunlit ocean waters?

(b) According to Figure 5.1(b), could fish living in the oceans at a depth of 1 000 m perceive: (i) moonlight; (ii) sunlight?

The illuminated zone in which light intensities are sufficient for photosynthetic primary production to lead to net growth of phytoplankton is called the **photic zone** (or euphotic zone). The greater the clarity of the water and the higher the Sun is in the sky, the greater the depth to which light penetrates and the greater the depth at which photosynthesis can proceed. The photic zone can be up to 200 m deep in clear waters of the open ocean, decreasing to about 40 m over continental shelves, and to as little as 15 m in some coastal waters. Only when the sea-bottom is shallow enough to be included in the photic zone are bottom-dwelling or **benthic** plants (e.g. attached seaweeds) able to grow – elsewhere in the ocean all plant life must float, i.e. it is **planktonic**. The wavelength of light is also important in photosynthesis (see Section 5.1.4).

Between the photic zone and the ocean floor is the **aphotic zone** where plants cannot survive for long, because light intensities are insufficient for photosynthetic production to meet the requirements of respiration (see also Section 6.1.3). Below about 1 000 m depth in the oceans, daylight can no longer be perceived (Figure 5.1, Question 5.1(b)). This means that throughout most of the oceans there is no external light at all. The only light is that provided by those fishes and other organisms that possess bioluminescent (light-producing) organs (and by human explorers using submersibles and other equipment). Note that the term *aphotic zone* is sometimes restricted to depths below about 1 000 m where daylight is absent, the region between this depth and the photic zone being called the *disphotic* (or *dysphotic*) *zone*.

5.1.1 ILLUMINATION AND VISION

In the photic zone and upper parts of the aphotic zone, objects in the sea are illuminated by sunlight (or moonlight), the intensity of which decreases

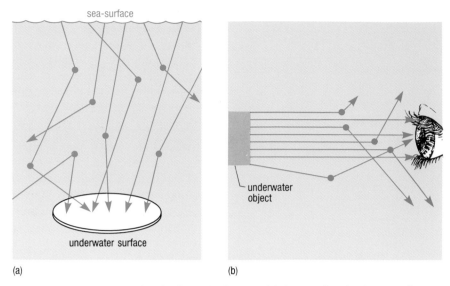

sea-surface

(a)

(b)

Figure 5.2 Diagrams showing the distinction between (a) the non-directional nature of illumination of an underwater surface by the downwelling irradiance and (b) the directional requirements of underwater vision – light *scattered* towards the eye cannot be focused to form part of a coherent image.

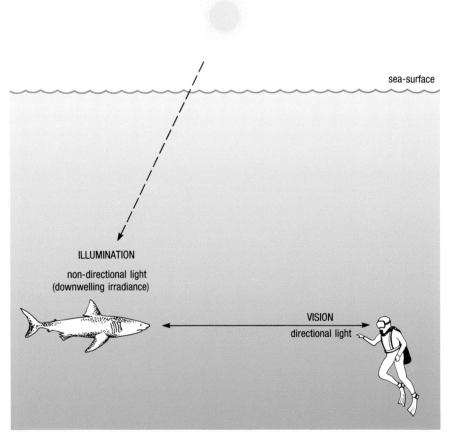

Figure 5.3 Illumination and vision under water. The more turbid the water, the greater the attenuation of light by absorption and scattering, the less the illumination at a particular depth, and the shorter the distance over which objects can be seen.

exponentially with depth from the surface, because it is attenuated by absorption and scattering (Figure 5.1). This so-called **downwelling irradiance** is diffuse, i.e. non-directional, because light illuminating an underwater object has not all taken the shortest path to it from the sea-surface; and light is scattered away from the object as well as towards it (Figure 5.2(a)). For an object to be *seen*, however, light emanating from the object must be directional, because a coherent image can only be formed if light travels directly from the object to the eye or camera (Figure 5.2(b)).

The distinction between illumination and vision is nicely illustrated in Figure 5.3. The fish is illuminated by non-directional light, but the image must be transmitted to the diver's eye by directional light for the fish to be seen. Perhaps a more everyday example is provided by a foggy day: your surroundings disappear but your view does not go black – in other words, you have illumination but no vision.

Looking at Figures 5.2 and 5.3, which do you think will be subject to the greater degree of attenuation: the non-directional light that provides illumination, or the directional light required to produce a coherent image in the eye or camera?

Light scattered away from an object being illuminated by the downwelling irradiance is 'compensated for' by light scattered towards the object. Light scattered out of the direct path from object to eye cannot be similarly 'compensated for' because light scattered towards the eye will not contribute to a coherent image, even though it originated from the object. So, it is the directional light associated with vision that is subject to greater attenuation.

5.1.2 UNDERWATER VISIBILITY: SEEING AND BEING SEEN

Visibility is a matter of contrast. An object may show up against its background either because it is a different colour, or because it has a different brightness (or both). Brightness contrast is more important than colour contrast in the marine environment, except in the upper few tens of metres of the photic zone (e.g. in the clear waters of the tropical reef environment, where colour contrasts are very important for inter- and intra-specific recognition, camouflage, deterring predators, and so on). At depths greater than a few tens of metres, the downwelling irradiance has not only been much attenuated by absorption and scattering, it has also become almost monochromatic, because of selective absorption of different wavelengths. Accordingly, at the low levels of light typical of most of the underwater world, even the eyes of animals that can normally distinguish colours must use the more sensitive night vision cells, with which everything is seen in shades of grey.

Contrast will decrease with distance, for two reasons: first, the light from the object being observed is attenuated by absorption and scattering; secondly, some of the incoming sunlight (or moonlight) is scattered towards the observer throughout the entire length of the path of sight. This effectively produces a 'veil of light', behind which the object becomes progressively more indistinct, until it disappears against its background.

QUESTION 5.2   Why do you think many fishes living in upper parts of the aphotic zone have dark upper surfaces but silvery undersides?

Figure 5.4   Examples of fishes with luminous organs. The upper seven fish represent a large number of species (some containing huge numbers of individuals) which inhabit the upper part of the aphotic zone (mainly the disphotic zone). These fish are quite small (sizes in the order of a few centimetres) and are shown here at roughly life size. Many species make diurnal vertical migrations, from depths in the order of 600 to 2 000 m by day to depths of 100 to 500 m at night. The preferred depths and precise extent of the migrations appear to be peculiar to each species. For example, the two hatchet-fish shown here (c and g) live at around 400–600 m depth and exhibit little vertical migration (although some other species of hatchet-fish do), whereas the lantern-fish (e) is found at depths of 650–1 700 m by day, but at night ascends to 50–300 m (and in cooler waters, to the surface).

The two lower species live at depths of 1 500–3 000 m and are much larger: they attain lengths of about 1.5 m. Both use a light organ as a lure for prey. The gulper eel (h) has a light organ on the tip of its tail, which it can recurve over its mouth. The angler fish has a luminous organ ('fishing lure') powered by light-emitting bacteria. (i) shows the adult female, with parasitic male (j) attached underneath – an adaptation to life in the depths, where finding a mate at the appropriate time in a sparsely populated environment could be difficult.

(a) Lightfish, *Vinciguerria attenuata*; (b) bristlemouth, *Cyclothone microdon*; (c) hatchet-fish, *Argyropelecus gigas*; (d) lantern-fish, *Myctophum punctatum*; (e) lantern-fish, *Lampanyctus elongatus*; (f) dragonfish, *Bathophilus longipinnis*; (g) hatchet-fish, *Argyropelecus affinis* (see photos); (h) gulper eel, *Eurypharynx pelecanoides*; (i) and (j) deep sea angler fish, *Ceratias holboelli* .

Hatchet-fish, *Argyropelecus affinis*, seen from below, showing line of photophores. Each contains a magenta-coloured filter which modifies the bioluminescent emission to match the spectral distribution of daylight in the sea.

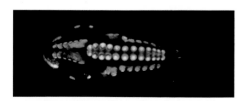

Bioluminescing *A. affinis* seen from below.

The light field becomes virtually symmetrical at depths of about 250 m, which means that the intensity of illumination is similar whether you look upwards or downwards. In the 250–750 m depth range, many fishes have silvered flanks, produced by interference 'mirrors' formed out of crystals of guanine (a nitrogenous compound), precisely orientated so as to function vertically when the fish is in its normal upright position. Light is reflected from these 'mirrors' with the same intensity as that of the background, thus effectively presenting zero contrast. Such fishes also have ventral photophores (luminous organs) which break up their silhouettes when viewed vertically from below; while their dorsal regions are black to minimize the contrast when viewed vertically from above – the hatchet-fish (*Argyropelecus*, Figure 5.4) is a common example.

In the upper parts of the aphotic zone – down to about 1 000 m and sometimes called the disphotic zone, Section 5.1 – where visual contact is still possible (Figure 5.1), many fish have developed large eyes to cope with the low light intensities. At greater depths, luminous organs arranged in distinctive patterns are developed in species that still depend on sight for contact, and fishes have become a uniform non-reflective black, so that they are not illuminated by the light of others.

In this environment, light is used in all the ways that colour is used in the terrestrial environment, e.g.:

- to deter would-be predators by appearing larger, e.g. with the aid of lights at the ends of long spines;
- to identify one's own species and/or mate:
- to provide signals whereby shoals can keep together;
- to break up the outline when viewed from below (see hatchet-fish, Figure 5.4, also many other fish species, as well as squid and some crustaceans); and
- as lures to attract prey, as well as headlamps to illuminate it (Figure 5.4).

5.1.3 MEASUREMENT

Instruments used for the measurement of underwater light fall into three main categories:

1   Beam transmissometers measure the attenuation of a parallel (collimated) light-beam from a source of known intensity, over a fixed distance. The ratio of light intensity at source and receiver (separated by a known distance) provides a direct measure of the attenuation coefficient for directional light, i.e. the percentage loss of light intensity (expressed as a decimal) per metre distance.

2   Irradiance meters accept light coming from any direction. The light is usually received by a teflon sphere or hemisphere, which measures ambient light downwelling from the surface – the downwelling irradiance. By making measurements of light intensity at different depths, the attenuation coefficient (called in this case the diffuse attenuation coefficient) for the non-directional downwelling irradiance can be determined. This is the appropriate coefficient for studies of photosynthetic primary production, because it relates to the exponential decrease in intensity of the downwelling irradiance, and hence to the depth of the photic zone.

QUESTION 5.3   Bearing in mind the earlier discussion of Figures 5.2 and 5.3:

(a)   Would you expect the attenuation coefficient for directional light to be greater or less than the diffuse attenuation coefficient for non-directional light?

(b)   For which of the two types of water represented in Figure 5.1(b) would these coefficients be greater, and for which would they be less?

As you might expect from Section 5.1.1, increased turbidity has a proportionately greater effect on directional than on non-directional light. The value of the ratio:

$$\frac{\text{attenuation coefficient} \quad (\text{directional light})}{\text{diffuse attenuation coefficient} \quad (\text{non-directional light})}$$

can be less than 3 in the open oceans, but as high as 10 or more in a turbid estuary.

3 Turbidity meters or *nephelometers* provide a direct measure of scattering in the water. A collimated beam illuminates a pre-determined volume of water which scatters light in all directions. The receiver is aimed at the centre of this scattering volume and can be rotated round it, so that variations in the scattering loss with direction relative to the light beam can be determined (Figure 5.5). As the degree of scattering is related to the amount of suspended material in the water, nephelometers provide a quantitative measure of turbidity, i.e. the concentration of suspended material. Nephelometry has been used, for example, to determine concentrations of suspended sediment in the deep ocean, and thus to provide information about the distribution and speed of bottom currents.

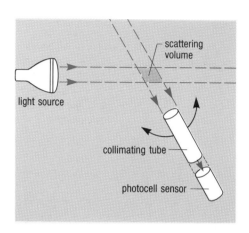

Figure 5.5 The principle of the nephelometer. The collimating tube in front of the photocell sensor can be aimed at the scattering volume from different directions.

The **Secchi disc** is a much more homely piece of equipment. It is simply a flat circular plate, 20–30 cm in diameter, either all white (Figure 5.6) or with two quadrants painted black and two painted white. It is lowered through the water column in a horizontal attitude until it is observed just to disappear. The depth at which this happens is called the **Secchi depth**, and it depends on the turbidity of the water. The Secchi disc is both cheap and easily made, and it has been used by oceanographers for over a century as a rapid means of assessing water clarity.

Simple empirical equations enable a good deal of information to be gleaned from the Secchi depth. The basic relationship for the vertically observed Secchi disc is:

$$Z_S = \frac{F}{C + K} \tag{5.1}$$

where:

Z_S is the Secchi depth;

C is the attenuation coefficient for directional light;

K is the diffuse attenuation coefficient for non-directional light (sometimes also called the extinction coefficient); and

F is a factor that depends on the reflectivity of the disc and that of the background, and the observer's own threshold perception of contrast. It is about 8.7 in clear oceanic water, but can be as little as 6 in turbid estuarine water.

QUESTION 5.4 (a) If the Secchi depth is 10 m, what is the sum of the attenuation and extinction coefficients $(C + K)$, assuming a value of 8 for the factor F?

(b) Which of the two coefficients contributes the larger proportion to the sum $(C + K)$, and is that contribution greater or less in clear or in turbid water?

Figure 5.6 The Secchi disc.

The reason why the Secchi depth provides a measure of the sum of these two coefficients is simply that the disc must be both *illuminated* (by downwelling irradiance to which the extinction coefficient is related), and *observed* (by directional light, to which the attenuation coefficient is related). Empirical relationships enable the Secchi disc to be used to estimate two useful parameters in the upper part of the water column:

$$\text{depth of photic zone} = 3Z_S \qquad\qquad (5.2)$$

$$\text{horizontal underwater visibility} = 0.7Z_S \qquad\qquad (5.3)$$

where visibility is defined as the distance at which the contrast of a black object becomes zero and it disappears against its background. Visibility relates to the attenuation coefficient, C, for directional light (equation 5.1) which, as noted above, is affected more than K by turbidity of the water (remember the analogy with a foggy day in Section 5.1.1). Underwater contrast, and hence visibility, depends also on the sighting angle: horizontal visibility is not necessarily the same as visibility looking upwards or downwards.

The numerical factors in equations 5.2 and 5.3 are likely to be different (by up to perhaps 15–20%) in different parts of the ocean. Finally, it is worth recording that neither temperature nor salinity of seawater has any appreciable effect on these phenomena: the coefficients C and K for clear seawater are virtually the same as those for pure water.

QUESTION 5.5 Would you expect Z_S to be greater or less (a) where primary production by phytoplankton is high or where it is low; (b) in coastal waters before or after a storm?

5.1.4 COLOUR IN THE SEA

Bearing in mind the information contained in Figure 2.5 and related text, can you suggest why many animals in the upper part of the aphotic zone (Section 5.1) have evolved black or red coloration?

Red animals appear red because they reflect red light, and the only light available from the downwelling irradiance in this 'twilight' zone is blue–green (Figure 2.5 shows that longer wavelengths of the visible spectrum have been absorbed at 100 m depth). So red animals will appear black (along with those that really are black) and will therefore be inconspicuous – an advantage for predator and prey alike.

The carotenoid pigments that provide the red colour also have maximum absorbence in the wavelengths emitted by most bioluminescent organs (photophores). This means that red fishes will not show up in the 'headlamps' of those that use light organs to illuminate their prey, such as *Diaphus* (a lantern-fish, cf. Figure 5.4(d) and (e)). Some fishes, however, have actually developed light organs that produce red light (e.g. *Pachystomasi*, a dragonfish, cf. Figure 5.4(f)) and their eyes contain the visual pigment to detect it. They can see without being seen, because red coloration is no camouflage when red light shines on it and the eyes of most other fishes are adapted to register only blue–green wavelengths.

As you read in Section 5.1, attenuation of underwater light results from a combination of absorption and scattering. Scattering of light by particles is largely independent of wavelength, but absorption is not. The principal absorbers in the sea, as listed in Section 5.1, absorb different wavelengths of light in different proportions.

(a) *Algae*: The most common plant pigment used in photosynthesis is chlorophyll-*a*, which strongly absorbs light at either end of the visible spectrum, i.e. red and blue-violet light. Light in the middle of the spectrum is reflected, which is why chlorophyll-*a* appears green. Many algae contain pigments that absorb light energy at other wavelengths, so light in the wavelength band from 400 nm (deep violet) to 700 nm (dark red) is described as *photosynthetically active radiation* (PAR).

Figure 5.7 contains similar information to Figure 2.5, but for a narrower wavelength band. It compares the energy spectra of solar radiation reaching different depths in various types of water. In brief, blue–green light (450–500 nm) penetrates furthest in the open ocean, and in fact about 35% of light of this wavelength that is incident on the surface reaches a depth of 10 m. In turbid coastal water, on the other hand, yellow–green light (500–550 nm) penetrates deepest, but only about 2% of that incident on the surface reaches a depth of 10 m.

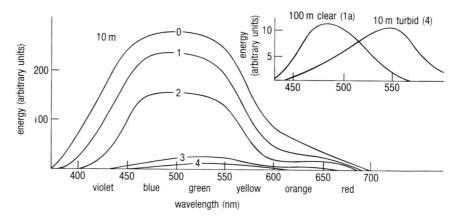

Figure 5.7   Energy spectra at a depth of 10 m for: pure water (0), clear oceanic water (1), average oceanic water (2), average coastal water (3), and turbid coastal water (4).
Inset: An energy spectrum at 100 m depth in clear oceanic water (1a) compared with that for 10 m in turbid coastal water (4). Compare this Figure with Figures 2.5 and 5.1 and note that it represents only a small part of the spectrum shown in Figure 2.5.

QUESTION 5.6   Does Figure 5.7 help to explain why seaweeds growing subtidally in clear coastal water (e.g. kelp) are very often brown?

(b) *Particulate matter*:   At normal concentrations, inorganic and organic particles other than algae absorb weakly but scatter strongly. Their comparatively small absorption is mainly in the blue range so their effect tends to be swamped by that of the dissolved organic compounds (see below).

(c) *Dissolved organic compounds*:   These are variously known as **yellow substances**, Gelbstoff, or gilvin. During the decomposition of plant tissue, organic material is broken down into CO_2, inorganic compounds of nitrogen, sulphur and phosphorus (the nutrients) and complex humic substances. It is these metabolic products that give some inland waters their distinctive yellow–brown coloration. They are brought to the sea by rivers, but are also produced in oceanic waters by the metabolism of plankton. Yellow substances absorb strongly at the short wavelength (blue) end of the spectrum, and reflect well (low absorption) in the yellow–red, hence the characteristic colour.

(d) *Water*: Water appears as a blue liquid, because absorption at the short wavelength (blue) end of the spectrum is relatively low whilst at the long wavelength (red) end it is high (Figure 2.5). Although water appears colourless in small quantities, its blue colour becomes apparent in clear tropical waters or a clean swimming pool. Absorption is so strong in the red that a 1 m-thick layer of pure water will absorb about 35% of incident light of wavelength 680 nm.

QUESTION 5.7   What percentage of incident red light is absorbed by 3 m of pure water?

Unproductive oceanic water carries little or no algae or yellow substances. It is therefore 'pure water blue' in colour. Blue is sometimes called the 'desert colour' of the oceans and is typical of many tropical waters. In recent years, a number of lakes in Scandinavia, Canada and elsewhere have 'died' (allegedly because of acid rain) and become a 'beautiful tropical blue'. In productive waters, red is absorbed by the water and blue is absorbed by yellow substances. This leaves 'sea-green' – the typical colour of productive mid-latitude waters.

There is commonly a colour change in the water across frontal boundaries (Section 4.4.3), especially where shelf water is separated from water of the open ocean.

In general, would you expect the spectral shift to be from blue to green or *vice versa*, in passing from the shelf to the deep water?

Shelf waters normally carry higher concentrations of yellow substances and suspended particles than the waters of the open ocean. So, we might expect the shift to be from green to blue when passing from the shelf to the deep water.

5.1.5 ELECTROMAGNETIC RADIATION AND REMOTE SENSING OF THE OCEANS

Passive remote sensing makes use of naturally reflected visible and near infrared wavelengths, as well as naturally emitted longer wavelength infrared and microwave radiation, to provide information about colour (and hence biological production and turbidity), temperature and ice cover at the surface of the oceans (e.g. Figures 1.5, 1.6, 2.3 and 4.15). It also provides information about surface roughness due to winds, waves, tides and currents, as well as about cloud type and extent, and the amount of water vapour in the atmosphere.

Active remote sensing involves the transmission of microwave pulses (radar) from aircraft or satellites, at wavelengths typically of a few cm, followed by measurement and analysis of the signals reflected by the surface. Imaging radar techniques provide information about sea-surface roughness (wave patterns and wave distribution) and ice cover. Radar has the advantage that it can penetrate clouds and is capable of providing high resolution.

We have seen that electromagnetic radiation can travel only short distances through water, so remote sensing and aerial photography provide direct information only about surface or near-surface waters, depending on wavelength; also, wave and ripple patterns can vary according to bathymetry, which can therefore sometimes be inferred from radar images. It follows also that radio communication is not possible under water, even though the attenuation coefficient for longer wavelength radio waves is less than it is for light. It is in fact possible to communicate with submarines submerged at depths of not more than a few tens of metres, either by using very long wavelength (very low frequency, VLF) radio waves, or laser beams from satellites. Laser light is very intense and in the 450–500 nm (blue–green) wavelength band it can penetrate far enough below the surface to be useful, before its energy is lost by attenuation. But that is the limit to which electromagnetic radiation can be used in the oceans. For both remote sensing and communication *within* the oceans, therefore, it is necessary to make use of much slower-moving acoustic radiation.

5.2   UNDERWATER SOUND

Although both light and sound can be considered to travel as waves, they are fundamentally different. As stated in Section 5.1, light is a form of electromagnetic energy. It propagates most effectively through a vacuum and in general less well as the density of material increases. Sound or acoustic energy involves the vibration of the actual material through which it passes and thus, in general, propagates best through solids and liquids, less well in gases and not at all in a vacuum.

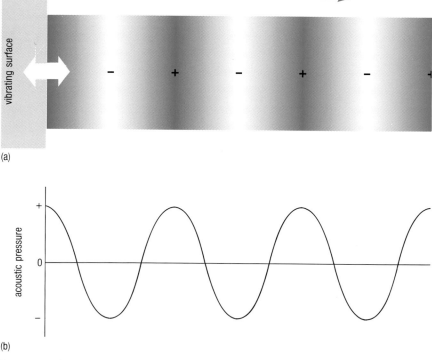

Figure 5.8   Characteristics of the acoustic wave. (a) Propagation of alternating zones of compression and rarefaction. (b) Sinusoidal rise and fall of acoustic pressure as the sound wave passes.

In short, sound is a form of pressure wave, propagated by vibrations that produce alternating zones of compression (molecules closer together) and rarefaction (molecules further apart) (Figure 5.8(a)). All sounds result from vibrations (e.g. the vibrating membrane of a loudspeaker or the vibrating sound organ of a deep-sea animal). Sound waves are thus not sinusoidal in the way that we normally consider wave motions to be. However, the *acoustic pressure* rises and falls in a sinusoidal manner as the sound wave passes (Figure 5.8(b)). So, as with other types of wave motion, sound waves can be characterized by their amplitude (a measure of intensity or loudness of the sound) and frequency (f) or wavelength (λ, lambda), which are related to speed (c) by the expression:

$$c = f\lambda \tag{5.4}$$

5.2.1 THE MAIN CHARACTERISTICS OF SOUND WAVES IN THE OCEANS

The wavelengths of acoustic energy that are of interest in the ocean range from about 50 m to 1 mm. Taking the velocity of sound in seawater as approximately 1 500 m s^{-1}, this corresponds to frequencies from 30 Hz* to 1.5 MHz. (For comparison, sound frequencies above about 20 kHz cannot be heard by the normal human ear.)

When acoustic energy is emitted uniformly in all directions by a point source in the middle of a homogeneous mass of seawater, it spreads outwards, producing spherical surfaces of constant pressure (remember that these are compressional waves), centred on the point source. The acoustic intensity will decrease with increasing distance from the source as a result of:

1 **Spreading loss** due to being spread over an increasingly greater surface area. The surface area of a sphere is proportional to the square of the radius of the sphere, and thus the spreading loss is proportional to the square of the distance travelled. Spreading loss is independent of frequency (see also Section 5.2.2). (Spherical spreading loss also occurs in the case of light, of course, but attenuation in water is so great over short distances that spreading loss is less important.)

2 **Attenuation** due to **absorption**, the conversion of acoustic energy into heat and chemical energy; and **scattering**, due to reflection by suspended particles and air bubbles. Scattering is largely independent of frequency; absorption is not. At high frequencies, viscous absorption predominates (i.e. absorption due to the viscosity of the water itself), and in freshwater this is the dominant cause of attenuation by absorption over much of the frequency range (Figure 5.9). However, in seawater at intermediate and low frequencies, the principal mechanism of absorption is dissociation of the $MgSO_4$ ion pair and of the $B(OH)_3$ complex (see Section 6.3.1). These split up into their constituent ions on the passage of a sound wave, and this process extracts energy from the sound wave – it is called 'relaxation' by acousticians. At very low frequencies (a few hundred Hz or less), it seems likely that the main cause of attenuation by absorption is inhomogeneities in the water column.

---

* The hertz (Hz) is the unit of frequency = 1 cycle per second. 1 kHz = 1 kilohertz = 10^3 Hz. 1 MHz = 1 megahertz = 10^6 Hz.

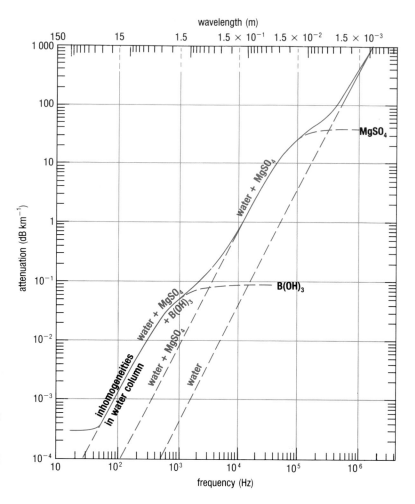

Figure 5.9 The attenuation of acoustic energy as a function of frequency in seawater, showing the dominant causes of attenuation and how they change according to frequency. The curves in this diagram are for a specific temperature and pressure; attenuation varies somewhat according to changing conditions. (dB = decibel, the unit of measurement for sound intensity.)

5.2.2 THE SPEED OF SOUND: REFRACTION AND SOUND CHANNELS

The speed, c, of compressional waves is given by

$$c = \sqrt{\frac{\text{axial modulus}}{\text{density}}} \qquad (5.5)$$

The axial modulus of a material is a measure of its elasticity in the context *both* of the ability to regain its original shape following compression, *and* of resistance to that compression; thus, the axial modulus of water is greater than that of air.

QUESTION 5.8 From equation 5.5, c varies inversely with density, implying that denser materials have lower acoustic velocities. This is seldom the case in naturally occurring materials. For instance, can you suggest how the speed of sound in water compares with that in air and in rock? Can you account for the apparent anomaly with the help of equation 5.5?

Both the axial modulus and the density of seawater depend on its temperature, salinity and pressure, and thus c becomes rather a complex function of these three variables in the ocean.

Raising the temperature of seawater lowers its density, and thus from equation 5.5 we should expect the speed of sound, c, to increase with increased water temperature. In the surface layers of the oceans, an increase in temperature of 1 °C leads to an increase in c of about 3 m s⁻¹.

We know that increased salinity leads to higher density, and so, at first sight, from equation 5.5 the speed of sound should decrease with increasing salinity. However, an increase in salinity also increases the axial modulus (the liquid becomes less compressible), and this more than counteracts the increase in density. For example, in surface layers of the oceans, an increase of 1 part per thousand in salinity actually results in an *increase* of about $1.1 \, \text{m s}^{-1}$ in c (the speed of sound in seawater is therefore greater than in freshwater – see Table 5.1).

Just as the speed of (seismic) sound waves increases with depth in the Earth, so the speed of acoustic waves increases with depth in the oceans (except in the sound channel, see below). The increase in axial modulus with pressure is greater than the corresponding increase in density, and so c becomes greater (equation 5.5). An increase in depth of 100 m will produce an increase in pressure of almost exactly 10 atmospheres ($10^6 \, \text{N m}^{-2}$, Figure 4.3), and the effect of this is to increase c by about $1.8 \, \text{m s}^{-1}$.

In the top few hundred metres of the ocean below the mixed surface layer, where temperature changes are large (Figures 2.6 to 2.8), c will be controlled mainly by temperature, and to a smaller degree by salinity and depth. Below the permanent thermocline, however, neither T nor S varies greatly, and so pressure becomes the dominant control on c.

A convenient empirical formula for the speed of sound in seawater over the temperature range 6 °C to 17 °C is:

$$c = 1\,410 + 4.21T - 0.037T^2 + 1.14S + 0.018d \qquad (5.6)$$

where T and S are, of course, temperature and salinity, and d is depth (in metres), to which pressure is directly proportional.

QUESTION 5.9   Calculate the speed of sound in seawater of temperature 10 °C and salinity 35, at 100 m depth.

An acoustic wave travelling vertically in the ocean will not be significantly affected by refraction because it is travelling essentially at right angles to the interfaces between layers of different density. However, a wave travelling

Figure 5.10   (a)   A typical temperature profile in the ocean.

(b)   A typical profile of the speed of sound in the ocean. The speed of sound is shown to increase with depth in the mixed layer, region I (see also Figure 5.11) because T and S are more or less constant there and so c is controlled chiefly by pressure. Region II coincides with the permanent thermocline (and main halocline), where c is controlled chiefly by changes in T and S. Below the permanent thermocline, in region III, the speed of sound is almost entirely controlled by pressure.

(c)   Idealized sketches illustrating refraction at interfaces where the speed of sound changes. (i) Upward refraction (regions I and III); and (ii) downward refraction (region II). From Snell's law:

$$\frac{c_{\text{greater}}}{c_{\text{less}}} = \frac{\sin i}{\sin r}$$

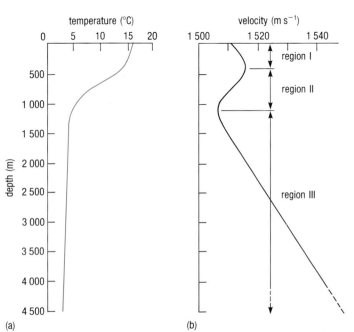

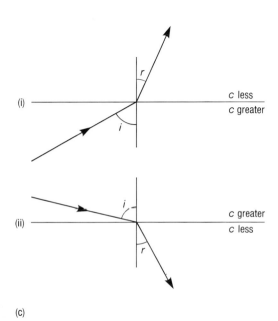

(a)   (b)   (c)

horizontally may be subject to considerable refraction because it will meet such interfaces at low angles. In regions I and III of Figure 5.10(b), a sound wave will be refracted upwards, because the speed of sound decreases upwards (cf. Figure 5.10(c)) whereas in region II it will be refracted downwards, because the speed of sound decreases downwards (cf. Figure 5.10(c)).

The paths that would be followed by acoustic waves may be determined from a knowledge of the values of c throughout the ocean, and ray diagrams can be drawn, as in Figure 5.11. The rays are simply lines drawn perpendicular to the propagating wave front, and they therefore represent the direction of propagation. Note that most rays are focused on the boundary between regions II and III, whereas there is a **shadow zone** in the vicinity of the boundary between regions I and II that is penetrated only by waves that have been reflected at the surface of the ocean. The channel in which rays are trapped by refraction at the boundary between regions II and III is known as the **sound channel**, which is in effect a 'wave guide' for sound in the oceans.

The spreading loss for energy emitted in the sound channel is proportional only to the distance travelled. This is because the energy is constrained by the sound channel to spread outwards mainly in the two horizontal dimensions. Therefore, the surfaces of constant acoustic pressure are cylindrical, not spherical (cf. Section 5.2.1, item 1), and the area of the curved surface of a cylinder is proportional to its radius (Figure 5.12). The information summarized in Figures 5.11 and 5.12 is of considerable significance in the use of acoustic energy in the oceans.

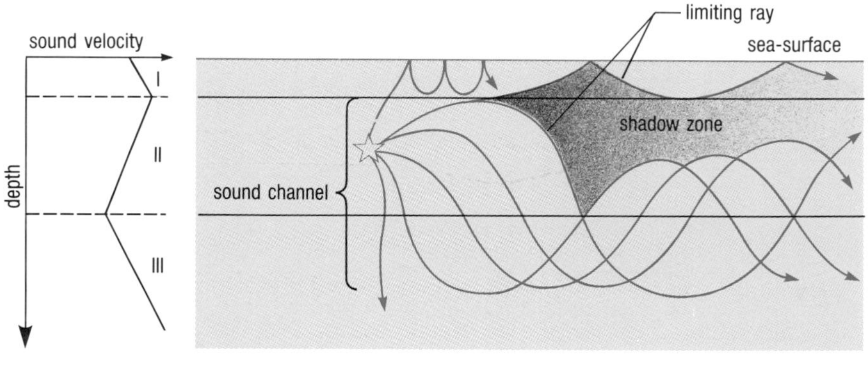

Figure 5.11   An example of a ray diagram for a sound emitted in region II of Figure 5.10(b), showing a sound channel and a shadow zone. (See text for further discussion.) The shadow zone is defined by the limiting rays, reflected at the sea-surface and/or refracted at the boundary between regions I and II.

Figure 5.12   Illustrations showing:

(a)   Spherical spreading loss from a point source. Surfaces of constant acoustic pressure are spherical and spreading loss is proportional to r^2.

(b)   Cylindrical spreading loss from a point source, as in the sound channel. Surfaces of constant acoustic pressure are cylindrical and spreading loss is proportional only to r.

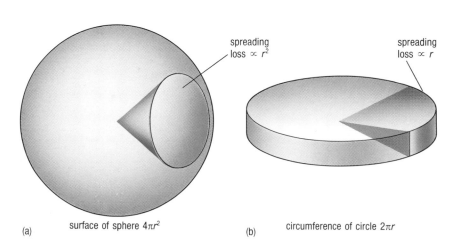

surface of sphere $4\pi r^2$ (a)

spreading loss $\propto r^2$

spreading loss $\propto r$

circumference of circle $2\pi r$ (b)

5.2.3 USES OF ACOUSTIC ENERGY IN THE OCEANS

The main disadvantage of using sound waves, in comparison with light waves, is their much greater wavelength (and lower frequency) which means that the resolution they can provide is much less; i.e. the smallest object that can be distinguished (about three wavelengths) is comparatively large. Frequency and wavelength are inversely proportional to one another (equation 5.4 and Figure 5.9): the higher the frequency, the shorter the wavelength, and *vice versa*. For maximum resolution of objects using underwater acoustic systems, therefore, the highest possible frequency is desirable. However, Figure 5.9 shows that attenuation is very frequency dependent.

Do higher or lower frequencies experience more attenuation in seawater?

Attenuation is greatest at high frequencies (short wavelengths) and least at low frequencies (long wavelengths). For example, losses from attenuation are about 5% per nautical mile (3% per km) at 5 kHz, increasing to about 90% per nautical mile (70% per km) at 30 kHz. So, to keep attenuation to a minimum, the lowest possible frequency must be used. However, we have just seen that for maximum resolution the highest possible frequency is desirable. Designers of acoustic systems used in the oceans therefore have to arrive at a compromise, according to whether range or resolution is more important.

Applications of acoustic energy in the oceans
There are four major categories:

1 *Passive acoustic systems*: These involve the use of receiving devices – hydrophones – to listen to the sounds that are present, such as those emitted by whales, fish, or submarines. Analysis of the frequency spectra of the 'sounds' will usually assist in identifying their sources.

The other three categories come under the general heading of *active acoustic systems*.

2 **Sonar** *(SOund Navigation And Ranging)*: An acoustic signal is emitted and reflections are received from objects within the water (perhaps fish or submarines) or from the sea-bed. When the acoustic wave travels vertically down to the sea-bed and back, the time taken will provide a measure of the depth of the water, if c is also known (either from direct measurement or from temperature, salinity and pressure data). This is the principle of the echo-sounder, which is now universally used on seagoing vessels, often as a navigation aid. A commercial echo-sounder may well have a beam width of 30–45° around the vertical, but for specialized applications (such as the detection of fish or submarines, or detailed studies of the sea-bed) beam widths of less than 5° are used and the direction of the beam can be varied. Note that although Figure 5.10 shows the effects of temperature, salinity and pressure on the speed of sound in seawater (c. 1 500 m s^{-1}) to be relatively small, even slight changes in c can lead to appreciable errors in depth measurement, and the degree of error may be increased by poor resolution.

Echo-sounding techniques for depth determination and sea-bed mapping have become very sophisticated, with the development of towed sonar devices, such as *SeaBeam* and *Hydrosweep,* which are multi-beam echo-sounding systems that determine the water depth along a swath of sea-floor beneath the towing ship, producing very detailed bathymetric maps.

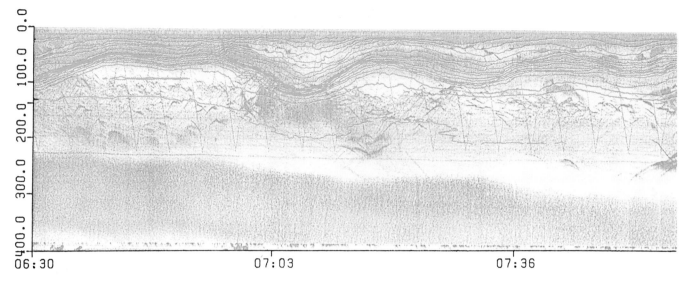

Figure 5.13   A typical sonogram (at 50 kHz) showing two scattering layers. The depth scale is in metres and the horizontal scale gives the time of day. The undulating upper band is in the thermocline, with the temperature structure (independently but simultaneously determined) superimposed as contours at 0.1 °C intervals. (The topmost contour is 10.9 °C.) Scattering in the thermocline could be due to backscatter caused by changes in acoustic impedance (see following text) associated with changes of temperature and density. Less coherent scattering below the thermocline is due to fish and zooplankton. The lower and more regular scattering layer (250–300 m depth) is believed to be due to the zooplankton *Meganactyphanes norvegica*, a species which undertakes diurnal vertical migrations. The regular zig-zag line is the trace of a conductivity–temperature–density (CTD) probe being used in 'yo-yo' mode.

Sidescan imaging systems, such as *GLORIA* (*G*eological *LO*ng *R*ange *I*nclined *A*sdic), *SeaMARC*, and *TOBI* (*T*owed *O*cean *B*ottom *I*nstrument) produce the equivalent of aerial photographs or radar images, using sound rather than light or microwaves to do so. Echo-sounding is also much used by fishermen, because even individual fish will produce an echo (see below), and shoals of fish or other animals can be identified as **scattering layers** within the water column (Figure 5.13).

Sonar has well-known military applications, especially in submarine warfare; and many marine animals have sonar-type mechanisms for the echo-location of prey or of other individuals in the group, as well as for identification and communication. Whales and dolphins are perhaps the best known for this ability – whales have been known to communicate with one another across entire oceans, using the sound channel. It is said that dolphins are also capable of stunning or even killing their prey with bursts of very intense acoustic energy; and that squid and octopus have evolved a simple protection against this form of attack – they are deaf.

Acoustic impedance is a measure of the acoustic behaviour of a material and determines how good a 'target' it will be for sonar systems:

$$\text{impedance}, Z = \rho c \tag{5.7}$$

So the acoustic impedance of seawater is about:

$$1.03 \times 10^3 \, \text{kg m}^{-3} \times 1\,500 \, \text{m s}^{-1} = 1.55 \times 10^6 \, \text{kg m}^{-2} \, \text{s}^{-1}$$

The reflection of acoustic energy can only occur at the interface between two media of different acoustic impedances. For reflection normal to the interface, the reflectivity R is given by:

$$R = \frac{Z_1 - Z_2}{Z_1 + Z_2} \times 100\% \tag{5.8}$$

where Z_1 and Z_2 are the acoustic impedances of the two materials on either side of the interface.

QUESTION 5.10   What is the reflectivity if $Z_1 = Z_2$?

Similarly, reflectivity will be at a maximum where $Z_1 - Z_2$ is greatest. Table 5.1 gives typical values of c, Z and R for some common materials.

Table 5.1 Acoustic properties of some common materials.

Material	Acoustic velocity c (m s^{-1})	Acoustic impedance $Z = \rho c$ ($\times 10^6$)	Reflectivity in seawater R (%)
Air (20 °C)	343	0.000 415	100
Freshwater (15 °C)	1481	1.48	–
Seawater (35‰, 15 °C)	1500	1.54	–
Wet fish flesh	~1450	1.6	1.9
Wet fish bone	~1700	2.5	24
Steel	6100	47	94
Brass	4700	40	92
Aluminium	6300	17	83
Perspex	2570	3.06	33
Rubber	1990	1.81	8
Concrete	3100	8	68
Granite	5925	16.0	82
Quartz	5750	15.3	82
Clay	~3000	7.7	67
Sandstone	~3300	~7.6	67
Basalt	~6000	~16.8	84

3 *Telemetry and tracking:* Locations may be identified and objects tracked in the oceans if they are equipped with acoustic transmitting devices. This is the basis of **Sofar** (*SO*und *F*ixing *A*nd *R*anging) technology, widely used for military purposes, such as the location of submarines, wrecked aircraft and sunken ships. Scientific use includes the charting of subsurface currents by means of floats equipped with acoustic sources. The density of *Sofar floats* can be adjusted so that they are neutrally buoyant at a particular specified depth (i.e. they sink to that depth and stay there, because their density is the same as that of the surrounding water), and they then drift passively in the prevailing current at that depth. If they emit their signals in the sound channel, they can be monitored by hydrophones perhaps thousands of kilometres away.

In addition to tracking the movement of water in the currents which move them along, transmissions from Sofar floats can also be used to pass other information. For example, if a temperature-sensing device is arranged to control either the frequency of the transmitted signal or the interval between successive signals, temperature data can also be provided.

The accuracy of Sofar fixes depends on reliable knowledge of the speed of sound in the oceans, especially within the sound channel. Figures 5.10 and 5.11 can be regarded as representing theoretical ideal situations. In practice, factors including seasonal and other fluctuations of temperature and salinity in time and space can lead to variations in the depth of the sound channel as well as losses from it. However, it is important to bear in mind that sound waves (rays) which 'leak' out of the sound channel tend to be reflected or refracted back into it (Figure 5.11): sound in the oceans travels with least loss in the sound channel, and Sofar devices work best within or near it. For this reason, the sound speed structure of the sound channel has been

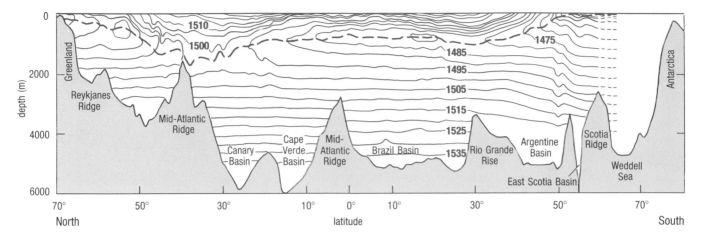

Figure 5.14   North–south section of the sound channel structure in the Atlantic along the 30.5° W meridian. Sound speeds are in m s⁻¹, and the approximate sound channel axis is indicated by a heavy blue broken line. Contours of equal speed are based on annual average data (near-surface structure above the axis in mid-latitudes is subject to seasonal variations). Note the increase in speed of sound both above and below the sound channel axis, cf. Figure 5.11.

mapped in some detail over most of the oceans, both by direct measurement and by computation using equations such as equation 5.6 with the many thousands of T and S measurements that have been made over the years. Figure 5.14 is a section resulting from one compilation of such data.

QUESTION 5.11   (a)   Is the speed of sound in the sound channel the same throughout Figure 5.14?

(b)   What can you say about variations in the depth of the axis of the sound channel with latitude?

(c)   Why does sound speed increase above the sound channel, where both temperature and salinity increase; and also below it, where temperature and salinity decrease?

4   *Current measurement*:   Sound may be used to measure current velocity by exploiting the *Doppler effect* whereby the measured sound frequency is affected by relative movement between an acoustic source and the point of measurement. Narrow-beam echo-sounders are aimed at a particular volume of water and the shift in frequency between the sound waves emitted by the hydrophones and those back-scattered by particles in the water is measured. This Doppler shift in frequency is proportional to the current speed, which can therefore be determined. Thanks to rapid technological advances, since the late 1980s it has become routine practice for ship-mounted acoustic Doppler current profilers (ADCPs) to be used for continuous measurement of current velocities to depths of several hundred metres, while the ship is actually in motion.

Acoustic noise
When specific acoustic signals are being emitted and then listened to, as in applications 2 to 4 above, all other acoustic energy in the ocean is considered as **noise** above which the required signal is to be heard (this is analogous to the way in which 'atmospherics' can swamp the weak signal from a distant radio source). Amplifying a weak signal in an attempt to enable it to be heard above the noise simply increases both signal and background noise; in addition, **reverberations** (multiple reflections by particles in the water as well as at the ocean boundaries) may become a severe problem. Some noise in an acoustic system may be caused by its own electrical circuits or the electrical system of the ship; the ship may also be the source of mechanical noise from engines and other equipment.

Ambient noise produced in the sea itself falls into two categories: physical and biological. Physical noise is mostly wind-induced and is in the audible frequency range (between about 10 and 10^4 Hz): it includes the sound of waves and bursting bubbles, rainfall, moving ice and sediment shifting on the sea-bed. Biological noise is produced by communicating whales and dolphins, by the activity of some crustaceans (e.g. snapping shrimps), and by certain fish.

Most of the biological noise produced and detected by marine animals is at very low frequencies, i.e. less than 50 Hz; the lateral line system of many fishes, for example, is highly sensitive to these low frequencies. Only those animals with specialized auditory receptors can use sound for communication – in some deep-living fish, those receptors are the swim-bladders that are normally used for buoyancy.

Acoustic oceanography

Since the early 1970s, there have been considerable advances in the application of acoustic techniques to the investigation of relatively short-term changes within and between water masses, on a variety of scales, from microstructure to fronts and eddies, and even up to basin-scale phenomena.

Acoustic tomography has been used to investigate fronts and to identify and track mesoscale eddies (Section 4.4.4). The method relies on the fact that individual eddies have temperatures different from those in the surrounding water – there are both warm and cold eddies. It follows that the speed of sound between an acoustic source and receiver will change if an eddy passes between them.

In practice, it is the 'travel time' of sound between source and receiver that is measured. Would the travel time increase or decrease if a cold eddy passed through?

A fall in temperature causes a decrease in the speed of sound (Section 5.2.2) – so the travel time would increase. A typical experiment involves a whole array of moored acoustic sources and receivers to monitor a 'volume' of ocean that may be from 300 to 1 000 km across. Travel times of acoustic pulses from each source must be measured at each receiver, so the data set is enormous and analysis of the results requires powerful computers.

Conventional ship-based measurements of temperature and current velocity must also be made in the area of interest. Acoustic travel times are affected not only by the properties of the water through which the sound travels, but also by the currents transporting that water. Clearly, currents travelling with the sound will reduce travel times, and those travelling against it will increase travel times.

QUESTION 5.12 (a) Temperature changes of a degree or two and salinity changes of 0.1 are typically encountered across eddy boundaries. How important is salinity in the context of acoustic tomography?

(b) Relatively low frequencies are used: around 250 Hz. By reference to Figure 5.9, (i) why is this preferable to using high frequencies, and (ii) how might the acoustic pulses also be used to distinguish between bodies of more and less well-mixed water?

During the early 1990s, oceanographers in the USA and Europe put forward a controversial proposal for a very large (ocean basin) scale acoustic experiment to detect and monitor the effects of global warming (Section 2.1) in the oceans. The idea is simplicity itself: to measure travel times of acoustic pulses transmitted via the sound channel from one or more sources in the Southern Ocean (Heard Island is a likely site) to receivers thousands of kilometres away in the northern Atlantic and Pacific Oceans.

How could that detect any effect of global warming?

The speed of sound in water depends on temperature (Section 5.2.2). Progressive warming on a regional or global scale should be detectable as a cumulative decrease in travel times between source(s) and receivers. The acoustic pulses must be of low frequency to minimize attenuation (cf. Question 5.12) and therein lies the principal reason for controversy. Some marine biologists claimed that high-volume (*c*. 190 decibels) low-frequency (60–90 Hz) acoustic pulses could endanger whales and other marine mammals that rely on sound for communication over large distances. This issue has not been satisfactorily resolved, but the sheer size of the experiment and the logistics of mounting it suggest that costs could be prohibitive.

5.3 SUMMARY OF CHAPTER 5

1   Light and all other forms of electromagnetic radiation travel at a speed of 3×10^8 m s^{-1} in a vacuum (about 2.2×10^8 m s^{-1} in seawater). Light travelling through water is subject to absorption and scattering, and its intensity decreases exponentially with distance from the source. Sunlight sufficient for photosynthesis cannot penetrate to more than about 200 m depth, and this defines the limit of the photic (or euphotic) zone, within which photosynthetic primary production can occur. The aphotic zone extends from the bottom of the photic zone to the sea-bed. Sunlight penetrates through only about the upper 1 000 m of the aphotic zone; below that, the oceans are permanently dark. The downwelling irradiance from sunlight or moonlight provides the non-directional (diffuse) light required for underwater illumination. Underwater vision requires directional light: light must travel direct from object to eye for a coherent image to be formed. Directional light is subject to greater attenuation than non-directional light.

2   Underwater visibility depends on contrast, which is a function partly of object brightness or reflectivity and partly of attenuation with distance. Below depths of a few tens of metres, underwater light becomes virtually monochromatic, so contrast is mostly a matter of differences of light intensity rather than of colour. In lower parts of the aphotic zone, where many fish have bioluminescent organs (photophores), light is used in the same way as colour is used on land – for inter- and intraspecific recognition, camouflage, deterring predators, and so on.

3   Beam transmissometers are used to determine the attenuation coefficient (C) of directional light, and irradiance meters are used to determine the diffuse attenuation coefficient (K) of the non-directional downwelling irradiance. Nephelometers measure scattering and can be used to determine concentrations of particulate matter in the water. The Secchi disc is a simple piece of equipment for measurement of water clarity. By applying simple empirical equations, the measurements can be used to estimate visibility, attenuation coefficients, and the depth of the photic zone.

4   Water preferentially absorbs longer wavelengths of the electromagnetic spectrum, which is why water appears blue. 'Yellow substances' and suspended particles absorb shorter wavelengths, so turbid water tends to look yellow, while productive ocean waters have the green colour of cholorophyll. In clear water, about 35% of incident blue–green light penetrates to 10 m depth. In turbid water, about 2% of yellow–green light penetrates to 10 m depth. Photosynthesis is inhibited in turbid waters.

5   Passive remote sensing of the oceans makes use of reflected and radiated visible, infrared and microwave radiation, to determine properties such as sea-surface temperature and water colour. Active remote sensing uses microwave imaging radar techniques to obtain information about the state of the sea-surface. Electromagnetic radiation cannot penetrate far through water, so remote sensing with the electromagnetic spectrum can provide direct information only about surface or near-surface waters, depending on wavelength; and radio communication is all but impossible below the surface of the ocean.

6   Sound travels much more slowly than light through water but can travel much further, and so is used for remote sensing and communication in the oceans. Frequencies of interest in the oceans lie approximately in the 30 Hz to 1.5 MHz range. Sound intensities decrease with distance from the source because of two processes: (a) spreading loss, due to being spread out over (i) the surface of a sphere (loss proportional to distance^2), or (ii) the surface of a cylinder (loss proportional to distance), as in the sound channel; and (b) attenuation, due to (i) absorption by the water and reactions involving its dissolved constituents, notably the dissociation of $B(OH)_3$ and $MgSO_4$ (attenuation increases as frequency increases, and high frequencies are very rapidly attenuated), and (ii) scattering, i.e. reflection by suspended particles.

7   The speed of sound in seawater, c, increases as the axial modulus of seawater increases, and decreases as the density increases; it is about 1 500 m s^{-1}. A temperature rise of 1 °C causes an increase of about 3 m s^{-1}. A salinity increase of 1 causes an increase of about 1.1 m s^{-1}. A pressure increase equivalent to an increase in depth of 100 m causes an increase of about 1.8 m s^{-1}. The speed of sound is at a minimum both at the surface and in the sound channel.

8   Sonar is used for depth determination, sea-bed mapping, and the location of objects, especially fish and submarines; many marine animals also make use of the technique. The reliability of echo-sounding depends partly upon the acoustic impedance: the higher the impedance contrast between water and the material of the object sought, the better the 'target' provided.

9   Sofar is used for longer-range location, and also for tracking, especially of neutrally buoyant acoustic floats within and near the sound channel. To fix the position of Sofar devices reliably, variations of the speed of sound throughout the oceans must be known as accurately as possible. The axis of the sound channel lies between about 0.5 and 1.5 km depth throughout most of the oceans, between the latitudes of about 60° N and S. Poleward of these latitudes there is no sound channel.

10   In any acoustic receiving system there is background noise due to ambient sounds emanating from instrumental, physical and biological sources; and reverberation due to multiple reflections – scattering – by particles, and at the ocean boundaries.

11   Acoustic oceanography experiments make use of the effect of temperature and other properties on the speed and attenuation of sound in seawater, to detect and monitor relatively short-term changes within and between water masses on scales ranging from microstructure to whole ocean basins.

Now try the following questions to consolidate your understanding of this Chapter.

QUESTION 5.13   Why are longer wavelengths of the electromagnetic spectrum missing on Figure 5.7?

QUESTION 5.14   Some fishes are better acoustic targets than suggested by the reflectivity of wet fish flesh in Table 5.1. Can you explain why this might be?

QUESTION 5.15   Would you select a high or low frequency sonar system for (a) fishing, (b) submarine detection?

QUESTION 5.16   Which of the following statements are true?

(a)   The depth of the photic zone is typically much greater near coasts than in the open ocean.

(b)   The waters of the Gulf Stream issuing from the Straits of Florida are highly productive (see the quotation in Section 4.4.3).

(c)   Equation 5.6 could not in general be used for water below the permanent thermocline.

(d)   The reflectivity of the air–sea interface in Table 5.1 should more accurately be given as 99.946%.

QUESTION 5.17   Examine Figure 5.15. Explain which curve is for the winter, and which is for the summer. What is the depth of the sound channel axis, and why are seasonal changes minimal at and below this depth?

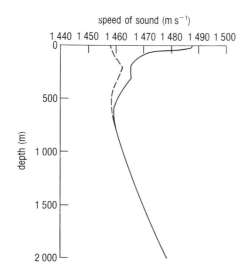

Figure 5.15   Variation of the speed of sound with depth at two seasons of the year. For use with Question 5.17.

CHAPTER 6   THE SEAWATER SOLUTION

> 'In my first science lesson we all watched intently round a pan of clear boiling salt water. When all the water had evaporated, we were awed at the sight of salt left at the bottom of the pan. Where had it come from?
>
> (Viki Capel, *New Scientist*, 1987.)

Up to now, we have mainly treated the salinity of seawater as a quantity approximating to 35 parts per thousand by weight. However, you have been introduced to the major ions that contribute 99.9 per cent of the salinity (Table 3.1), and in Section 3.1 we established the important principle of the constancy of composition of seawater. In this Chapter, we consider seawater as an aqueous solution and examine the sources and the behaviour of some individual constituents.

6.1   THE GROSS CHEMICAL COMPOSITION OF SEAWATER

Most of the 92 naturally occurring elements have been measured or detected in seawater, and the remainder are likely to be found as more sensitive analytical techniques become available. The elements so far determined show a vast range of concentrations, as you can see from Table 6.1.

QUESTION 6.1   There are major differences in the figures for sulphur, carbon and boron in Table 6.1, compared with Table 3.1. Why is that? Why would the use of molar concentrations avoid such disparities?

Particulate matter
There is a wide variety of suspended particles in seawater (the seston, Section 5.1), and the distinction between what constitutes material truly in solution (i.e. dissolved in the water) and what is particulate matter (i.e. in suspension) can present problems in the determination of the concentrations of some elements in seawater. A widely used procedure for separating dissolved from particulate fractions is filtration through a membrane having pores of diameter $0.45\,\mu m$. For most constituents this provides a satisfactory separation between dissolved and particulate matter, but for some it is less clear-cut. For example, iron in seawater occurs in hydrated forms such as $Fe(OH)_2$ or $Fe(OH)_3$. These tend to coalesce to form **colloidal** particles, which are so small that they remain in suspension indefinitely, unless some process occurs to aggregate them into particles large enough to settle under gravity. Thus, for iron there is a spectrum of sizes, ranging from true solution, through colloidal particles, to aggregated particles. Use of a membrane having pores of diameter $0.45\,\mu m$ therefore effects a purely arbitrary separation between dissolved and particulate fractions. The measured ratio of dissolved to particulate iron in a given sample can be increased or decreased simply by changing the pore size of the membrane, or by increasing the filtration pressure, which may break up the aggregates mechanically. This problem does not arise with all elements that occur in hydrated forms, however. In the case of $Al(OH)_3$ and $Si(OH)_4$, for example, filtration can satisfactorily distinguish dissolved from particulate fractions.

Table 6.1 Average abundances of chemical elements in seawater.

Element		Concentration (mg l^{-1}) (i.e. parts per million, p.p.m.)	Some probable dissolved chemical species	Total amount in the oceans (tonnes)
Chlorine	Cl	1.95×10^4	Cl$^-$	2.57×10^{16}
Sodium	Na	1.077×10^4	Na$^+$	1.42×10^{16}
Magnesium	Mg	1.290×10^3	Mg^{2+}, MgSO$_4$, MgCO$_3$	1.71×10^{15}
Sulphur	S	9.05×10^2	SO$_4^{2-}$, NaSO$_4^+$	1.2×10^{15}
Calcium	Ca	4.12×10^2	Ca^{2+}	5.45×10^{14}
Potassium	K	3.80×10^2	K$^+$	5.02×10^{14}
Bromine	Br	67	Br$^-$	8.86×10^{13}
Carbon	C	28	HCO$_3^-$, CO$_3^{2-}$, CO$_2$ gas	3.7×10^{13}
Nitrogen	N	11.5	N$_2$ gas, NO$_3^-$, NH$_4^+$	1.5×10^{13}
Strontium	Sr	8	Sr^{2+}	1.06×10^{13}
Oxygen	O	6	O$_2$ gas	7.93×10^{12}
Boron	B	4.4	B(OH)$_3$, B(OH)$_4^-$, H$_2$BO$_3^-$	5.82×10^{12}
Silicon	Si	2	Si(OH)$_4$	2.64×10^{12}
Fluorine	F	1.3	F$^-$, MgF$^+$	1.72×10^{12}
Argon	Ar	0.43	Ar gas	5.68×10^{11}
Lithium	Li	0.18	Li$^+$	2.38×10^{11}
Rubidium	Rb	0.12	Rb$^+$	1.59×10^{11}
Phosphorus	P	6×10^{-2}	HPO$_4^{2-}$, PO$_4^{3-}$, H$_2$PO$_4^-$	7.93×10^{10}
Iodine	I	6×10^{-2}	IO$_3^-$, I$^-$	7.93×10^{10}
Barium	Ba	2×10^{-2}	Ba^{2+}	2.64×10^{10}
Molybdenum	Mo	1×10^{-2}	MoO$_4^{2-}$	1.32×10^{10}
Uranium	U	3.2×10^{-3}	UO$_2$(CO$_3$)$_2^{4-}$	4.23×10^9
Vanadium	V	2×10^{-3}	H$_2$VO$_4^-$, HVO$_4^{2-}$	2.64×10^9
Arsenic	As	2×10^{-3}	HAsO$_4^{2-}$, H$_2$AsO$_4^-$	2.64×10^9
Titanium	Ti	1×10^{-3}	Ti(OH)$_4$	1.32×10^9
Zinc	Zn	5×10^{-4}	ZnOH$^+$, Zn^{2+}, ZnCO$_3$	6.61×10^8
Nickel	Ni	4.8×10^{-4}	Ni^{2+}, NiCO$_3$, NiCl$^+$	6.35×10^8
Aluminium	Al	4×10^{-4}	Al(OH)$_4^-$	5.29×10^8
Caesium	Cs	4×10^{-4}	Cs$^+$	5.29×10^8
Chromium	Cr	3×10^{-4}	Cr(OH)$_3$, CrO$_4^{2-}$, NaCrO$_4^-$	3.97×10^8
Antimony	Sb	2×10^{-4}	Sb(OH)$_6^-$	2.64×10^8
Krypton	Kr	2×10^{-4}	Kr gas	2.64×10^8
Selenium	Se	2×10^{-4}	SeO$_3^{2-}$, SeO$_4^{2-}$	2.64×10^8
Neon	Ne	1.2×10^{-4}	Ne gas	1.59×10^8
Cadmium	Cd	1×10^{-4}	CdCl$_2$	1.32×10^8
Copper	Cu	1×10^{-4}	CuCO$_3$, Cu(OH)$^+$, Cu^{2+}	1.32×10^8
Tungsten	W	1×10^{-4}	WO$_4^{2-}$	1.32×10^8
Iron	Fe	5.5×10^{-5}	Fe(OH)$_2^+$, Fe(OH)$_4^-$	7.27×10^7
Xenon	Xe	5×10^{-5}	Xe gas	6.61×10^7
Manganese	Mn	3×10^{-5}	Mn^{2+}, MnCl$^+$	3.97×10^7
Zirconium	Zr	3×10^{-5}	Zr(OH)$_4$	3.97×10^7
Niobium	Nb	1×10^{-5}	Nb(OH)$_6^-$	1.32×10^7
Thallium	Tl	1×10^{-5}	Tl$^+$	1.32×10^7
Thorium	Th	1×10^{-5}	Th(OH)$_4$	1.32×10^7
Hafnium	Hf	7×10^{-6}	Hf(OH)$_5^-$	9.25×10^6

Element		Concentration ($mg\,l^{-1}$) (i.e. parts per million, p.p.m.)	Some probable dissolved chemical species	Total amount in the oceans (tonnes)
Helium	He	6.8×10^{-6}	He gas	8.99×10^{6}
Germanium	Ge	5×10^{-6}	$Ge(OH)_4$, $H_3GeO_4^-$	6.61×10^{6}
Rhenium	Re	4×10^{-6}	ReO_4^-	5.29×10^{6}
Cobalt	Co	3×10^{-6}	Co^{2+}	3.97×10^{6}
Lanthanum	La	3×10^{-6}	$La(OH)_3$, La^{3+}, $LaCO_3^+$	3.97×10^{6}
Neodymium	Nd	3×10^{-6}	$Nd(OH)_3$, $NdCO_3^+$, Nd^{3+}	3.97×10^{6}
Cerium	Ce	2×10^{-6}	$Ce(OH)_3$, $CeCO_3^+$, Ce^{3+}	2.64×10^{6}
Lead	Pb	2×10^{-6}	$PbCO_3$, $Pb(CO_3)_2^{2-}$, Pb^{2+}	2.64×10^{6}
Silver	Ag	2×10^{-6}	$AgCl_2^-$	2.64×10^{6}
Gallium	Ga	2×10^{-6}	$Ga(OH)_4^-$	2.64×10^{6}
Tantalum	Ta	2×10^{-6}	$Ta(OH)_5$	2.64×10^{6}
Yttrium	Y	1×10^{-6}	YCO_3^+, Y^{3+}	1.32×10^{6}
Mercury	Hg	1×10^{-6}	$HgCl_4^{2-}$, $HgCl_2$	1.32×10^{6}
Dysprosium	Dy	9×10^{-7}	$Dy(OH)_3$, $DyCO_3^+$, Dy^{3+}	1.19×10^{6}
Erbium	Er	8×10^{-7}	$Er(OH)_3$, $ErCO_3^+$, Er^{3+}	1.06×10^{6}
Ytterbium	Yb	8×10^{-7}	$Yb(OH)_3$, $YbCO_3^+$	1.06×10^{6}
Gadolinium	Gd	7×10^{-7}	$Gd(OH)_3$, $GdCO_3^+$, Gd^{3+}	9.25×10^{5}
Praseodymium	Pr	6×10^{-7}	$Pr(OH)_3$, $PrCO_3^+$, Pr^{3+}	7.93×10^{5}
Samarium	Sm	6×10^{-7}	$Sm(OH)_3$, $SmCO_3^+$, Sm^{3+}	7.93×10^{5}
Tin	Sn	6×10^{-7}	$SnO(OH)_3^-$	7.93×10^{5}
Scandium	Sc	6×10^{-7}	$Sc(OH)_3$	7.93×10^{5}
Holmium	Ho	3×10^{-7}	$Ho(OH)_3$, $HoCO_3^+$, Ho^{3+}	3.97×10^{5}
Beryllium	Be	2×10^{-7}	$BeOH^+$	2.64×10^{5}
Lutetium	Lu	2×10^{-7}	$Lu(OH)^{2+}$, $LuCO_3^+$	2.64×10^{5}
Europium	Eu	2×10^{-7}	$Eu(OH)_3$, $EuCo_3^+$, Eu^{3+}	2.64×10^{5}
Indium	In	2×10^{-7}	$In(OH)_2^+$, $In(OH)_3$	2.64×10^{5}
Thulium	Tm	2×10^{-7}	$Tm(OH)_3$, $TmCO_3$, Tm^{3+}	2.64×10^{5}
Terbium	Tb	1×10^{-7}	$Tb(OH)_3$, $TbCO_3^+$, Tb^{3+}	1.32×10^{5}
Palladium	Pd	5×10^{-8}	Pd^{2+}, $PdCl^+$	6.61×10^{4}
Gold	Au	2×10^{-8}	$AuCl_2^-$	2.64×10^{4}
Bismuth	Bi	2×10^{-8}	BiO^+, $Bi(OH)_2^+$	2.64×10^{4}
Tellurium	Te	1×10^{-8}	$Te(OH)_6$	1.32×10^{4}
Radium	Ra	7×10^{-11}	Ra^{2+}	92.5
Protactinium	Pa	5×10^{-11}	Not known	66.1
Radon	Rn	6×10^{-16}	Rn gas	7.93×10^{-4}
Polonium	Po	5×10^{-16}	Po_3^{2-}, $Po(OH)_2$?	6.61×10^{-4}

IMPORTANT NOTES

1   Table 6.1 does not represent the last word on seawater composition. Even for the more abundant constituents, compilations from different sources differ in detail (cf. Note to Table 3.1). For the rarer elements, many of the entries in Table 6.1 will be subject to revision, as analytical methods improve. Moreover, most constituents behave non-conservatively (Section 4.3.4), making averages less meaningful.

2   Concentrations in Table 6.1 are by weight (p.p.m., i.e. $mg\,l^{-1}$ or $mg\,kg^{-1}$, because in practice it can generally be assumed that 1 litre of seawater weighs 1 kg). While this is convenient for some purposes, for many others it is more useful to express concentrations in molar terms. There are 6×10^{23} atoms in a mole (Avogadro's number). One mole of any element (or compound) has a mass in grams equal to the atomic (or molecular) mass of the element (or compound). Thus, a mole of calcium weighs 40 g; a mole of magnesium weighs 24 g; a mole of carbonate ion (CO_3^{2-}) weighs $12 + (16 \times 3) = 60$ g; and so on. See also Question 6.1.

The density of particulate matter is typically greater than that of seawater, so it tends to sink. However, the small size of most particles means that they can remain in suspension almost indefinitely.

The classical equation for the settling velocity, v, of a spherical object in a fluid medium is (Stokes' law):

$$v = \frac{1}{18} g \frac{(\rho_1 - \rho_2)}{\mu} d^2 \tag{6.1}$$

where:

g is the gravitational acceleration (m s^{-2})

d is the diameter of the particle (m)

ρ_1 is the density of the particle (kg m^{-3})

ρ_2 is the density of the fluid (kg m^{-3})

μ is the molecular viscosity of the fluid (N s m^{-2}) (cf. Table 1.1),

and the velocity, v, will be in m s^{-1}.

Equation 6.1 gives a first-approximation value for the speed at which seston particles sink in seawater.

QUESTION 6.2   The majority of particles making up the seston have diameters less than 2 μm. Assume an average density 1.5 times that of seawater, take values of the density and viscosity of water from Tables 1.1 and 1.2, and use 9.8 m s^{-2} for g. Use equation 6.1 to calculate v for a particle that is 2 μm in diameter. About how long would such particle take to sink 1 m?

The long period of time implied by Question 6.2 would be even greater if the particle were not spherical. In addition, it would be considerably extended by turbulence in the water column, which counteracts the tendency of particles to settle out of suspension. Equation 6.1 is also an (indirect) expression of the fact that the smaller a particle is, the greater will be the ratio of its surface area to its volume, and the greater the frictional resistance as it sinks through the water column. In order for sedimentary particles to reach the sea-bed from the surface in a reasonable period of time (say a month), they must be much larger than particles typical of the seston. (In fact, equation 6.1 is valid only for spherical particles with diameters less than about 100 μm. For particles with diameters greater than about 2 mm, the settling velocity is proportional to $d^{1/2}$ and equation 6.1 has a different form. For particles in the size range 100 μm to 2 mm, settling velocity is proportional to d^n, where $2 > n > \frac{1}{2}$.)

The major sources of particulate material in the oceans are:

1   Rivers, carrying particles in suspension to the sea, where the coarser fractions are deposited as sands, silts and clays.

2   Wind-borne (aeolian) dust (e.g. fine particles of quartz, clay minerals and iron oxide, silicified grass cells, freshwater diatom skeletons, as well as organic detritus, see below); in addition, volcanic ash particles, micrometeorites (cosmic dust), and material derived from the break-up of larger meteorites are continually supplied to the oceans by atmospheric fall-out. Much of this input rapidly sinks to the sea-bed, because most of it has densities in the range 2–3×10^3 kg m^{-3}, but some of the particles are small enough to contribute to the seston. Most inorganic particulate matter in the seston of the open ocean is probably of aeolian origin.

3   Biogenic particulate matter, i.e. particles resulting from primary and secondary biological production, comprising skeletal remains, faecal pellets and dead plant and animal matter (detritus). Much of this material has particle sizes of 100 µm or more, and sinks relatively quickly even though its density is not much greater than that of seawater. Smaller particles of organic matter (algal cells and detritus less than about 10 µm) sink very slowly (cf. Question 6.2), and they tend to accumulate near the top of the pycnocline.

Why should this happen?

Principally because turbulence in the mixed layer tends to prevent the particles from sinking, and the pycnocline is the base of the mixed layer. Also, the density of seawater increases rapidly with depth in the pycnocline (Figure 4.5), so the density *contrast* between particles and seawater decreases. In other words, the $(\rho_1 - \rho_2)$ term in equation 6.1 becomes smaller, and so does the settling velocity, v.

6.1.1 CLASSIFICATION OF DISSOLVED CONSTITUENTS

Major constituents of seawater are those that occur in concentrations greater than about 1 part per million (p.p.m, cf. Table 6.1) by weight, and they account for over 99.9% of the dissolved salts in the oceans. The major constituents of seawater are conventionally taken to be those listed in Table 3.1. Despite their relatively high concentrations, nitrogen and oxygen and silica (Table 6.1) are not generally included, because the first two are dissolved gases and the third is a nutrient (see Section 6.1.2); oxygen and silica in particular are strongly non-conservative and their concentrations vary greatly.

Minor and trace constituents make up the remainder of the elements in seawater. Although the distinction between the two is somewhat ill-defined, trace constituents are those with concentrations of about 1 part per billion (1 in 10^9, or 10^{-3} p.p.m.) by weight, or less. On that basis, elements below about titanium in Table 6.1 are trace constituents.

The oceanic distribution of individual major constituents is in general closely related to that of total salinity, because of the constancy of composition of seawater (Section 3.1). Most of these major constituents behave conservatively (Section 4.3.4) – the exceptions, calcium (Ca^{2+}) and carbon in its various forms, along with silica (SiO_2), are dealt with in the next Section. On the other hand, most minor and trace constituents behave non-conservatively, being affected by biological and chemical processes in which they are added to or removed from solution.

6.1.2 THE NUTRIENTS

Carbon is a fundamental requirement for the support of life anywhere on Earth. Because of the predominance of carbon dioxide among the dissolved gases (see Section 6.1.3), carbon forms the eighth most abundant dissolved element in ocean water (Table 6.1). The availability of dissolved carbon is therefore not generally considered to be a limiting factor in biological production, so it is not classified as a nutrient. More important constraints are the intensity of illumination, supply of oxygen, and the availability of nutrients, especially fixed nitrogen, chiefly as nitrate (NO_3^-, also as ammonium (NH_4^+, see below)), phosphorus as phosphate (PO_4^{3-}) and silicon as silica (SiO_2), sometimes also called silicate. Nutrients are utilized by **phytoplankton** (Figure 6.1(d)) – plant cells ranging in size from one to a few hundred microns – which drift in the surface waters of the oceans and photosynthesize carbohydrates from carbon dioxide and water.

Why can phytoplankton not grow below depths of about 100–200 m?

Light is essential for photosynthesis, and you have seen in Section 5.1 that phytoplankton can grow only in the photic zone, which is rarely deeper than 200 m and generally much less. It is in the photic zone, therefore, that nutrients are most heavily utilized. Phytoplankton form the base of oceanic food chains and nutrients move along the chains as grazing and predation take place. They are recycled (returned to solution) within the water column by excretion and microbial breakdown of organic particulate matter (detritus). The return of nutrients to solution by decomposition of organic matter is known as *re-mineralization*. Sinking of larger biogenic particles (faeces and corpses) and the vertical movements of **zooplankton** (Figure 6.1(e)) and other animals feeding on phytoplankton and detritus combine to cause a progressive downward movement of nutrients out of the photic zone. As a result, concentration profiles for nitrate, phosphate and silica typically look like those in Figure 6.1. On profiles such as those in Figure 6.1, the slope representing increased concentrations of nutrients with depth below the mixed surface layer is commonly called the **nutricline**.

The photic zone is continually being depleted of nutrients, and photosynthetic primary production will be inhibited unless there is vertical mixing, or vertical advection of nutrient-rich water from greater depths (*upwelling*). The term **biolimiting constituents** is sometimes applied to those nutrients whose availability in surface waters limits biological

Figure 6.1 Typical concentration profiles in subtropical and tropical waters for (a) phosphate, (b) nitrate, and (c) silica. Note that concentrations are in mol l^{-1}. (d) Living phytoplankton, mainly diatoms and dinoflagellates; the field of view is *c.* 1.75 mm across. (e) Living zooplankton, including copepods (planktonic crustaceans) and the planktonic larvae of various animals; the field of view is *c.* 1.75 cm across.

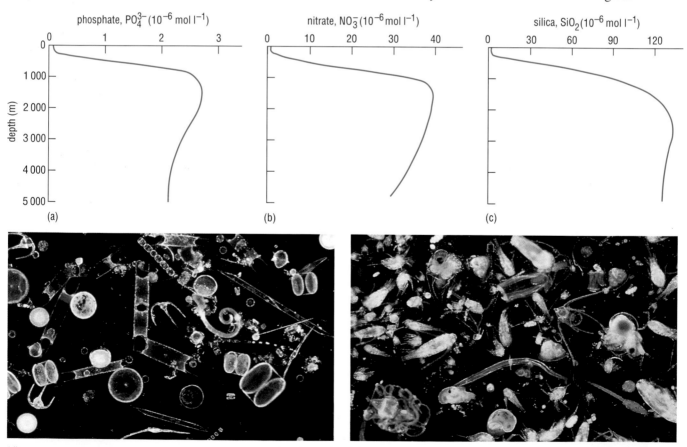

(a) (b) (c)

(d) (e)

production. They include nitrate, phosphate and silica (or silicate), and their characteristic profiles, showing almost total depletion in the mixed surface layer (Figure 6.1), are controlled principally by biological processes.

Nitrate and phosphate are used to form the soft tissues of organisms and the ratio of the molar concentration of nitrate and phosphate (the molar ratio) in ocean water is close to 15 : 1 for organic tissue (cf. Figure 6.1(a) and (b)); thus, when all the dissolved phosphate in surface waters has been used up, so has all the dissolved nitrate. Why nitrate and phosphate should occur in seawater in the same ratio that organisms require them remains one of the intriguing mysteries of seawater chemistry. There is still no answer to the question of whether organisms evolved to use the 15 : 1 molar ratio of N : P because it was there, or whether marine organisms themselves established the ratio through time.

Nitrogen, nitrate and ammonia

It is essential to recognize the crucial distinction between nitrogen as N_2 gas and fixed (combined) nitrogen as nitrate, NO_3^-, the nutrient. Dissolved nitrogen gas (N_2) is used hardly at all in marine biological processes, because only a few phytoplankton (principally the *Cyanobacteria*, or blue–green algae) are capable of fixing it, i.e. converting it into the organic nitrogen compounds required for growth. Average oceanic water contains about 9 ml l^{-1} (parts per thousand by volume) of nitrogen gas, N_2 (see Section 6.1.3 and Figure 6.4), which is equivalent to about 11 mg l^{-1} (p.p.m. by weight). The total concentration of nitrogen in average ocean water is given as 11.5 p.p.m. in Table 6.1; so, only a very small fraction can be in a form other than N_2 gas dissolved from the atmosphere. This small fraction is chiefly nitrate, NO_3^-, derived from the decomposition of organic matter on land (where most nitrogen-fixing bacteria occur) and supplied to the oceans by rivers. Rainfall also supplies small amounts of nitrate, produced mainly through the combination of atmospheric nitrogen and oxygen in lightning discharges, but partly also as fallout (washout) from industrial pollution – indeed, in some nearshore regions acid rain is a significant source of nitrate.

Ammonia, NH$_3$: Formed during bacterial decomposition of organic matter and excreted by zooplankton, ammonia occurs in solution in seawater chiefly as the ammonium ion, NH_4^+. It is eventually oxidized to nitrate, but it can also be used as a nutrient by phytoplankton (and it is the only source of nitrogen for many free-living bacteria), i.e. it is fixed nitrogen that is recycled within the photic zone (see also Section 6.3.5), and so it is the principal source of fixed nitrogen in nutrient-poor surface waters. Organic matter is continually sinking out of the photic zone, however, and ammonium concentration is generally low. Biological production can only be sustained if nitrate continues to be supplied to surface waters by rivers, atmospheric fallout, or upwelling from below the nutricline (Figure 6.1).

Silica

The third nutrient, silica (or silicate), is used to build the skeletons of planktonic plants (diatoms), and animals (radiolarians) (Figure 6.2). The silica secreted by organisms is an amorphous form and it is hydrated, so its formula is commonly written as $SiO_2.nH_2O$ (and it is sometimes called opaline silica or opal), but for brevity we shall use SiO_2 for both solid and dissolved silica. After the organisms die or are consumed, the skeletal debris sinks through the water column and slowly dissolves in deep water, giving concentration profiles like that in Figure 6.1(c).

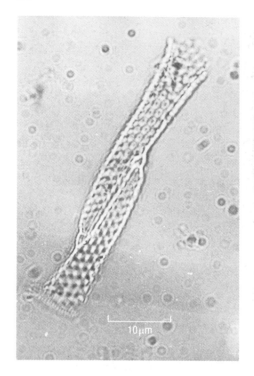

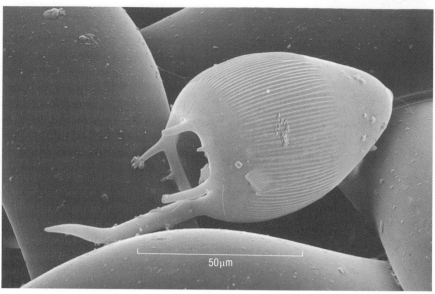

(b)   Radiolarian (zooplankton) skeleton also of $SiO_2.nH_2O$. Radiolaria are plentiful in deep-sea sediments (hence the name radiolarian ooze), commonly accompanied by diatoms. The background is part of the sampling net.

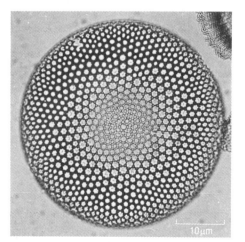

Figure 6.2   (a)   Skeletons of diatoms (phytoplankton) formed of opaline silica ($SiO_2.nH_2O$). Diatoms are common in deep-sea deposits under areas of high surface biological productivity. See also Figure 6.1(d).

QUESTION 6.3   Why are well-stratified surface waters likely to be more rapidly depleted in nutrients than a well-mixed upper water column?

Many marine plants and animals form skeletons of calcium carbonate, $CaCO_3$ (Figure 6.3), so carbon is used for both the soft and hard parts of organisms. The biological utilization of carbon and calcium in the marine environment is a major component in the global cycles of these two elements. Both are abundant in the seawater solution, however (Table 6.1), and the amounts used by organisms are small in relation to the total abundance (cf. Section 3.1). The dissolved species of elements such as carbon and calcium are sometimes called **bio-intermediate constituents** because although they show some depletion in surface waters, they are never exhausted, even in regions of very high biological production.

Constituents whose concentrations in solution are unaffected by biological activity are sometimes called **bio-unlimited constituents** – and these are also the constituents that behave conservatively in seawater (e.g. sodium and chloride).

As noted at the end of Section 4.3.4, in the context of dissolved constituents the terms *conservative* and *non-conservative* reflect the extent to which the concentration of a constituent is affected by chemical or biological processes *in relation to its overall concentration in seawater*. For example, sodium takes part in many biological processes, but its oceanic abundance is so great that these processes have a negligible effect on its concentration and it is classified as a conservative constituent.

6.1.3   DISSOLVED GASES

Three-quarters of the mass of the atmosphere is concentrated in the lowest 10 km, and this part of the atmosphere shows no variations in the proportion of its major constituents: nitrogen (78%), oxygen (21%) and argon (1%).

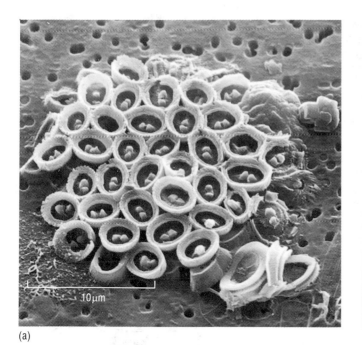

(a)

(b)

(c)

Figure 6.3 (a) Coccolithophore remains, skeletons of the phytoplankton *Discosphaera* sp. and *Coronosphaera* sp., formed of $CaCO_3$. The coccosphere disaggregates during deposition and is rarely preserved intact. The component platelets (the coccoliths) are common in deep-sea sediments and in chalk deposits (e.g. the White Cliffs of Dover). (b) Skeleton (test) of foraminiferan zooplankton (*Guembelitria* sp.) formed of $CaCO_3$. Foraminiferan remains are another common constituent of deep-sea sediments. (c) Skeletons (tests) of pteropods (zooplankton), formed of aragonite, $CaCO_3$ – see later text. These are molluscs (gastropods), easily recognizable in sediments because of their relatively large size and elongate conical shape.

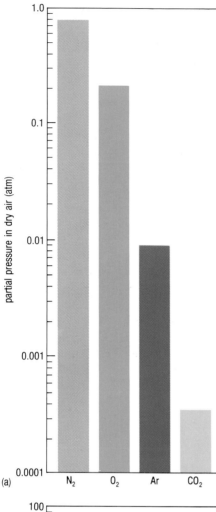

(a)

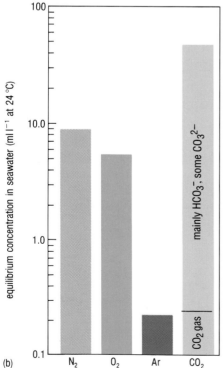

(b)

Concentrations of atmospheric gases are conventionally given by volume. Carbon dioxide accounts for only about 0.035% of the total (Figure 6.4(a)).

An important feature of Figure 6.4(a) is that the vertical scale is given in **partial pressure**, which is identical with percentage composition by volume: for example, if you took away all the gases except oxygen, the 21% oxygen would give a pressure of 0.21 atmosphere. Equilibrium concentrations of the four most abundant gases in seawater at 24 °C are shown in Figure 6.4(b).

QUESTION 6.4 What is the ratio of nitrogen to oxygen (i) in the atmosphere and (ii) in seawater? How much more or less soluble is nitrogen than oxygen?

Note that although concentrations of gases are given in $mg\,l^{-1}$ (p.p.m. by weight) in Table 6.1, these are numerically not very different from the volumetric concentrations ($ml\,l^{-1}$) for oxygen, nitrogen and argon in Figure 6.4. That is because the densities of these gases are respectively 1.43, 1.23 and $1.77\,kg\,m^{-3}$ (so, to a first approximation, $1\,m^3$ weighs $1\,kg$, 1 litre weighs 1 gram, and 1 ml weighs 1 mg).

Figure 6.4(b) shows that the solubility of CO_2 in seawater is many times greater than that of nitrogen and oxygen. That is because dissolved CO_2 reacts with water, forming carbonic acid and its dissociation products bicarbonate and carbonate:

$$CO_2\,(gas) + H_2O \rightleftharpoons H_2CO_3(aq) \rightleftharpoons H^+(aq) + HCO_3^-(aq) \qquad (6.2)$$

$$H^+(aq) + CO_3^{2-}(aq)$$

where (aq) means aqueous, i.e. in solution.

Because of reaction 6.2, CO_2 as dissolved *gas* is present only in very small amounts in seawater: $0.23\,ml\,l^{-1}$ at 24 °C and atmospheric pressure. That is roughly one two-hundredth part of the total shown in Figure 6.4(b); the remainder is mostly in the form of bicarbonate ion (HCO_3^-) and some carbonate ion (CO_3^{2-}), a distribution we return to in Section 6.3.2.

The information in Figure 6.4 is a useful starting point for the discussion of dissolved gases, but it must be treated with caution, as the data apply to seawater at one temperature and pressure only. *The solubility of gases generally decreases with increasing temperature and salinity, and increases with increasing pressure.* In addition, Figure 6.4(b) is based upon the assumption that there is equilibrium between atmosphere and ocean across the air–sea interface. At equilibrium, rates of gaseous diffusion are the same in both directions (i.e. there is no net flux of gas into or out of seawater) because the number of molecules of gas entering the seawater solution is equalled by the number of molecules escaping back to the atmosphere. This is probably valid to a first approximation for the four most abundant gases (Figure 6.4), but not for many that occur in much lower concentrations, some of which are discussed later (see Table 6.2).

The distribution of gases at deeper levels in the oceans is achieved mainly by currents and by turbulent rather than molecular diffusion. Downward redistribution is slow, however, and the oceans may take many hundreds of

Figure 6.4 (a) Partial pressure (= volume proportions) of the four most abundant gases in the atmosphere, together totalling more than 99.9% of the atmosphere, the rest being made up of minor gases. (b) Equilibrium concentrations by volume of these four gases dissolved in seawater at 24 °C as controlled by the solubilities of the gases at their atmospheric partial pressures. Note different logarithmic scales on vertical axes. For discussion of CO_2, see text.

years to cquilibrate with the overlying atmosphere – in other words, it takes a long time for the effects of changes in the processes that control gas exchange at the air–sea interface to be 'transmitted' throughout the oceans.

Biological activity plays an important role in redistribution of oxygen and carbon dioxide below the surface, and largely determines the form of their concentration profiles.

Oxygen

Ocean surface waters are consistently supersaturated with oxygen (Figure 6.5), partly due to liberation of oxygen during photosynthesis, but mainly as a result of air bubbles formed at the crests of waves being forced down into the water column, where part of the gas they contain is driven into solution by the increased hydrostatic pressure.

Near the bottom of the photic zone, there is a balance between the amount of carbon that phytoplankton fix by photosynthesis and the amount they 'burn' or dissipate (oxidize) in respiration. The depth at which this balance occurs is called the **compensation depth**, and it can also be defined as the depth at which the amount of oxygen produced by phytoplankton during photosynthesis over a 24-hour period equals the amount they consume in respiration. In short, for a phytoplankton population at the compensation depth, reaction 6.3 has reached a balance: for every mole of carbon fixed (or mole of oxygen released) in photosynthesis, a corresponding mole of carbon is oxidized (using up a mole of oxygen) in respiration:

$$CO_2(gas) + H_2O \underset{\text{metabolic energy (respiration)}}{\overset{\text{light energy (photosynthesis)}}{\rightleftharpoons}} \underset{\text{organic matter}}{(CH_2O)n} + O_2(gas) \quad (6.3)$$

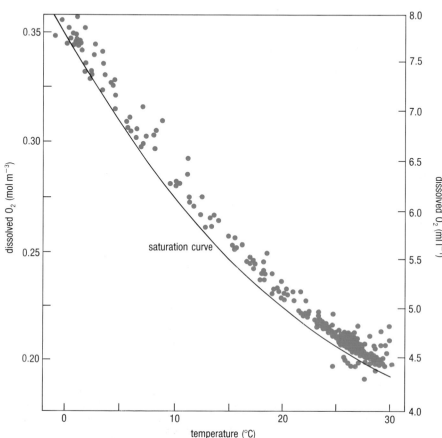

Figure 6.5   The saturation curve for oxygen (solid black line) and measured concentrations (blue dots) in ocean surface waters, determined as part of the GEOSECS programme.

Photosynthesis does not cease at the compensation depth, but below it light intensities are insufficient for net phytoplankton growth to occur (Figure 5.1); i.e. below the compensation depth more carbon is dissipated in plant respiration than can be fixed by photosynthesis. Algae can survive to considerable depths – viable plant cells have been recovered from depths of several thousand metres – but they cannot actually grow once they sink below the compensation depth. For practical purposes, then, the compensation depth can be taken to represent the lower limit of the photic zone.

At greater depths, oxygen continues to be consumed in the respiration of both animals and plants and in the microbial decomposition (oxidation) of organic detritus. But it is not being replenished, partly because downward mixing and diffusion of dissolved oxygen from the surface is slow; and partly because photosynthesis declines to negligible levels below the photic zone. An **oxygen minimum layer** develops where abstraction of dissolved oxygen is greatest relative to the rate of replenishment, generally between 500 and 1 000 m depth. In some areas, such as the northern Indian Ocean and the eastern tropical Pacific (Figure 6.6), the water at these depths is very oxygen-deficient, and in extreme cases it can become completely anoxic. At greater depths in the open oceans, oxygen levels rise again, because of the input of cold, dense oxygenated water sinking in polar regions (cf. Figure 2.13). The vertical distribution of oxygen varies considerably from place to place (Figure 6.6), but in general it is almost a mirror image of those for phosphate and nitrate. The top of the oxygen minimum layer (e.g. Figure 6.6) is just below the base of the mixed surface layer, where sinking organic detritus tends to accumulate, as you read at the end of Section 6.1.

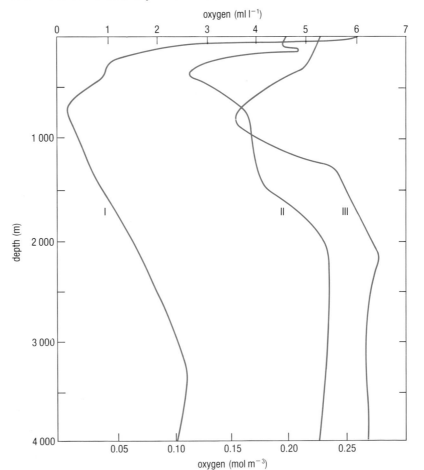

Figure 6.6   The vertical distribution of dissolved oxygen (concentration in ml l^{-1} and mol m^{-3}, i.e. 10^3 mol l^{-1}) I: South of California; II: eastern part of the South Atlantic; III: Gulf Stream. Water in the oxygen minimum layer of profile I is sub-oxic, almost anoxic.

Where strong oxygen minimum layers coincide with the sea-floor, anoxic sediments may be deposited. This typically occurs along continental margins, where there is high biological productivity. In addition, where basins are isolated from oxygenated deep-sea circulation by shallow barriers (sills), e.g. in the Black Sea, anoxic sediments form at all depths below the level at which oxygen is exhausted. The term **oxic** is commonly applied to the well-oxygenated water that characterizes most of the world's oceans. Water in oxygen minimum layers (Figure 6.6) is often described as *sub-oxic* and, as noted above, *anoxic* conditions occur where oxygen is absent.

QUESTION 6.5   (a)   Why is water sinking in polar regions more highly oxygenated than that elsewhere?

(b)   On what grounds might we infer a greater degree of biological activity to have occurred in the water column represented by profile I in Figure 6.6 than in the other two?

Carbon dioxide

Just as in the case of oxygen, the lower the temperature, the more CO_2 goes into solution. Below the thermocline, however, where temperature is virtually constant, the solubility of carbon dioxide becomes almost entirely a function of pressure: increased pressure forces more CO_2 into solution to form carbonic acid and its dissociation products (reaction 6.2). This example of Le Chatelier's principle is well known to consumers of bottled or canned beverages that froth or fizz upon being opened: when the pressure is released reaction 6.2 goes to the left and CO_2 gas escapes. However, even at very high pressures in the deep ocean (Figure 4.3), the proportion of CO_2 as dissolved *gas* is still less than 2% of the total amount of CO_2 in solution. The rest is represented chiefly by bicarbonate (HCO_3^-) and carbonate (CO_3^{2-}) ions, cf. Figure 6.4(b).

pH of seawater: Seawater in equilibrium with atmospheric CO_2 is slightly alkaline, with a pH of around 8.1–8.3. The pH may rise slightly through the rapid abstraction of CO_2 from surface waters during photosynthesis (reaction 6.3), but it does not normally exceed 8.4 except in tidal pools, lagoons and estuaries. We have seen that below the photic zone, the CO_2 absorbed in photosynthesis is exceeded by the CO_2 released in respiration. As CO_2 concentrations increase, so pH falls, typically to values of about 7.7–7.8. It can reach values of 7.5 or less in waters of reduced salinity or in anaerobic (anoxic) conditions, where bacteria using reduction of sulphate as a source of oxygen for the decomposition of organic matter release H_2S into solution (cf. Section 3.1.1). However, when the sulphate has been used up, decomposition of organic matter under anaerobic conditions involves the reduction of CO_2 itself, and leads to the formation of hydrocarbons, such as methane, CH_4. Under these conditions, the pH may rise to values as high as 12. We look at pH again in Section 6.3.2.

Some minor gases

All but one of the gases listed in Table 6.2 (overleaf) are produced by organisms in surface waters. The surface waters are supersaturated with these gases, so they must escape to the atmosphere; in other words, their net flux is from sea to air. The exception in Table 6.2 is *sulphur dioxide*, whose net flux is from air to sea. Its sources include volcanism and industry (fossil fuel burning and metal smelting), and oxidation of natural organic sulphur compounds (including dimethyl sulphide, see below). In the atmosphere, it is oxidized to SO_3 gas, which quickly combines with water to form sulphuric acid aerosols which contribute to the problem of acid rain. Along with dust and aerosols from volcanic gases and sea salt (Section 2.2.1), sulphate aerosols from oxidation of sulphur compounds provide nuclei for condensation of atmospheric water vapour into clouds and rain. So, much of the air to sea flux may be as sulphate ions (SO_4^{2-}) as well as gaseous SO_2.

Table 6.2 World-wide sea–air fluxes for some gases.

Gas	Total oceanic flux (g yr^{-1})	Direction of net flux
Sulphur dioxide, SO_2	1.5×10^{14}	air $\rightarrow$ sea
Nitrous oxide, N_2O	1.2×10^{14}	sea $\rightarrow$ air
Carbon monoxide, CO	4.3×10^{13}	sea $\rightarrow$ air
Methane, CH_4	3.2×10^{12}	sea $\rightarrow$ air
Methyl iodide, CH_3I	2.7×10^{11}	sea $\rightarrow$ air
Dimethyl sulphide, $(CH_3)_2S$	4.0×10^{13}	sea $\rightarrow$ air

Surface seawater is generally oversaturated with *nitrous oxide* (N_2O) because of bacterial activity, and the resulting flux from sea to air may be important in the oceanic nitrogen budget. The rate at which fixed nitrogen enters the oceans from river inflow and rain is about 8×10^{13} gN yr^{-1} (chiefly as dissolved nitrate, NO_3^-). About 10% of this (9×10^{12} gN yr^{-1}) is removed to marine sediments as organic nitrogen compounds in undecomposed organic detritus, leaving 90% to be accounted for.

QUESTION 6.6 What is the sea→air flux of N_2O given in Table 6.2, in terms of gN yr^{-1}? Use relative atomic masses N = 14, O = 16. Does this make up the balance of the nitrogen input to the oceans that is not removed to the sediments?

Carbon monoxide (CO) and *methane* (CH_4) provide an interesting contrast. Their concentrations in surface seawater are similar, but the atmospheric concentration of carbon monoxide is much *less* than that of methane. The concentration gradient across the interface is therefore much greater for CO than for CH_4, which accounts for the order of magnitude difference in their fluxes (Table 6.2). The sea $\rightarrow$ air fluxes of these gases are not important components of the global carbon budget. Both gases are produced by microbial breakdown of organic matter (some CO is also produced by algal respiration), and both are oxidized to CO_2 in the atmosphere. Methane is particularly interesting because anaerobic (anoxic) reducing conditions are required for its formation. Methane supersaturation in typically well-oxygenated (oxic) surface seawater (Figure 6.5) would therefore seem paradoxical, were it not that methane-generating bacteria have been found in anoxic micro-environments within faecal pellets and other particulate organic detritus sinking out of surface waters.

Methyl iodide (CH_3I) and *dimethyl sulphide, DMS* (($CH_3)_2S$) are also unstable in oxygenated environments. They are produced by some phytoplankton species near the sea-surface and persist long enough to pass into the atmosphere, where they are decomposed, dimethyl sulphide being oxidized to sulphate as outlined above.

6.1.4 DISSOLVED GASES AS TRACERS

Whatever the present position of a subsurface body of water, it must at some time have been at the surface, where diffusion across the air–sea interface will have determined its dissolved gas content. Once the water has sunk away from the surface and become isolated from the atmosphere, the concentrations of dissolved gases will change as a result partly of mixing and partly of biological or other reactions.

Oxygen is used as a tracer because it is abundant, biologically important and easily measured. The longer a water mass is isolated from the atmosphere, the lower its oxygen content becomes. By tracking back along the concentration gradient of oxygen, the source region of the water mass can be located, and the changes that have gone on within the water mass since its isolation from the surface can be inferred.

Figure 6.7(a) shows a high concentration of dissolved oxygen in surface waters at high latitudes in the Atlantic Ocean. This diminishes gradually with depth and distance towards the Equator, consistent with the sinking of water masses in polar regions. Figure 6.7(b) shows relatively high concentrations of dissolved oxygen in surface waters in the South Pacific. These diminish northwards and with depth, consistent with the sinking of Antarctic waters.

Figure 6.7   Sections showing dissolved oxygen (ml l⁻¹) in (a) the western Atlantic (cf. Figures 2.6 and 3.3) and (b) the Pacific (along about 170° W).

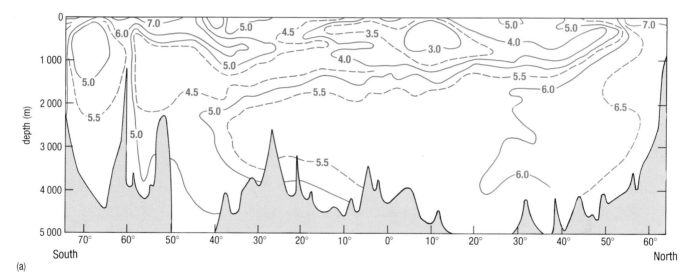

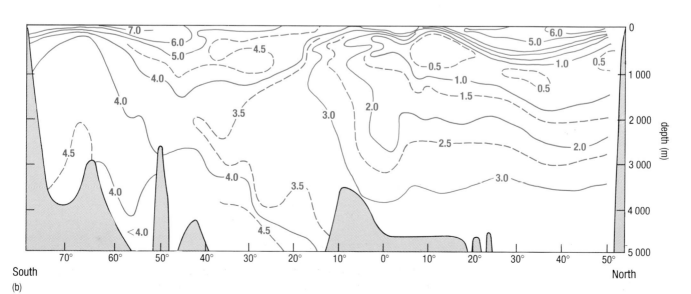

QUESTION 6.7 (a) Why is dissolved oxygen a non-conservative constituent, and why does its concentration progressively decrease with time after the water in which it is dissolved has left the surface?

(b) Use the distribution of dissolved oxygen in Figure 6.7(a) to identify the main subsurface Atlantic water masses: North Atlantic Deep Water, Antarctic Intermediate Water, and Antarctic Bottom Water (see also Figures 2.6, 3.3 and A1 (with the answer to Question 4.1)).

(c) How does the distribution of dissolved oxygen in the North Pacific in Figure 6.7(b) suggest the absence of a source region of deep water there?

6.2 SOURCES AND SINKS, OR WHY THE SEA IS SALT

In this and subsequent Sections we shall look at the ways in which supply and removal of dissolved constituents contribute to the overall compositional balance of seawater, and then look at some of the chemical reactions in which the different constituents participate.

If rivers are the chief source of the dissolved salts in seawater, why is seawater not simply a concentrated version of the average composition of all rivers?

As outlined at the start of Chapter 3, the answer lies in the chemical behaviour of dissolved constituents as they circulate through the hydrological cycle. The next few Sections develop this theme in more detail.

6.2.1 COMPARISON OF SEAWATER WITH OTHER NATURAL WATERS

Figure 6.8 shows the average concentrations of the principal dissolved constituents in rainwater, river water and seawater. The averages of rainwater and river water conceal considerable variations, but the basic pattern is the same all over the world.

QUESTION 6.8 (a) How many times more dilute than seawater is (i) rainwater, (ii) river water?

(b) Is rainwater or river water closer to seawater in composition?

Thus, to change rainwater into river water clearly requires the addition of substantial amounts of certain constituents, and these are provided mainly by the chemical weathering of rocks. Rainwater contains dissolved gases, particularly CO_2 and SO_2, both of which form acidic solutions in water, so that rainwater is a weak acid (pH 5.7). When rain falls on the land, the acidity is neutralized by reaction with minerals in soils and rocks:

$$CaCO_3(s) + CO_2(gas) + H_2O \rightarrow Ca^{2+}(aq) + 2HCO_3^-(aq) \tag{6.4}$$

(calcite, a common   (from rainwater)   (in solution)
mineral in
sedimentary rocks)

$$2NaAlSi_3O_8(s) + 2CO_2(gas) + 3H_2O \rightarrow Al_2Si_2O_5(OH)_4(s) + 2Na^+(aq) + 2HCO_3^-(aq) + 4SiO_2(aq,s)$$

(albite, a common   (from rainwater)   (kaolinite,   (in solution)   (silica, partly
mineral in igneous and   a clay mineral)   in solution)
metamorphic rocks)

$$\tag{6.5}$$

where (s) stands for solid.

The two representative examples shown in reactions 6.4 and 6.5 simplify the real situation, but they account broadly for the processes by which rainwater is transformed into river water. The exceptionally large increases in the concentrations of Ca^{2+} and HCO_3^- between rainwater and river water (Figure 6.8) arise from the fact that these ions can be produced from weathering both of carbonates (reaction 6.4), and of calcium-bearing silicates (e.g. reaction 6.5).

6.2.2 SEAWATER AND RIVER WATER

Seawater contains about 300 times more dissolved salts than average river water (Question 6.8), and a glance at Figure 6.8 shows the mix of elements dissolved in river water to be very different from that in seawater. In the marine environment, substantial amounts of HCO_3^-, Ca^{2+} and SiO_2 in particular must be removed from solution. We have established that some of the dissolved constituents in river water come from chemical weathering of

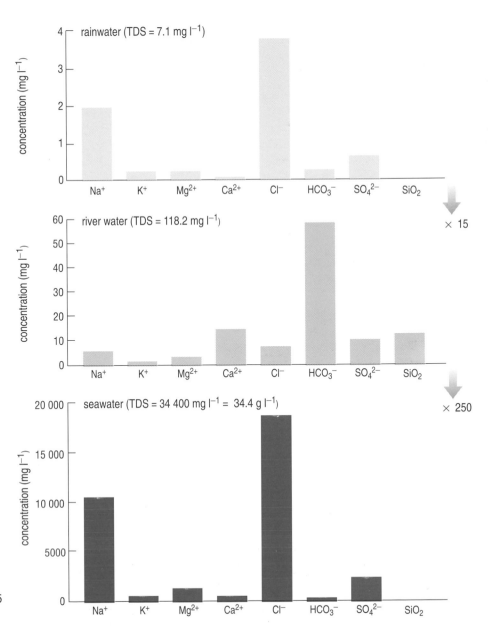

Figure 6.8   The average chemical composition of seawater, river water and rainwater for up to eight dissolved constituents, some at concentrations too low to appear (TDS = total dissolved salts). Note that total concentrations increase from rainwater to river water to seawater, also the scale change (arrows) of ×15 from (a) to (b) and ×250 from (b) to (c).

Figure 6.9 Marine aerosols (darkest blue arrow) transport seawater and its dissolved constituents into the atmosphere, from which they are removed by rain (pale blue arrow). Rivers entering the sea (medium blue arrow) thus contain not only dissolved constituents derived from rock weathering, but also recycled seawater constituents (cyclic salts). Compare Figures 6.8 and 6.10.

surface rocks. The remainder are recycled from the oceans via aerosols and rainfall (Figure 6.9, cf. Section 2.2.1). We shall now try to quantify the relative contributions from these two sources.

The average chloride content of continental crustal rocks is in the order of 0.01%, and so only a very minute proportion of the chloride in river water comes from weathering. It follows that virtually all the chloride in river water (Figure 6.8) must be from sea salt recycled via oceanic aerosols (Figure 6.9). This enables us to 'correct' river water compositions for recycled or **cyclic salts**.

How might we do this?

We can do it by applying the constancy of composition of seawater for major constituents (Section 3.1). Our basic assumptions are that all the chloride in river water is recycled from the oceans by rain (and snow), and that the other constituents are recycled in the same proportions as they occur in seawater. These assumptions have been applied in Figure 6.10 to 'correct' the measured concentrations in river water by subtracting the contributions from cyclic salts. What remains is the contribution from weathering. Note the complete disappearance of Cl^- in Figure 6.10 as a result of this correction.

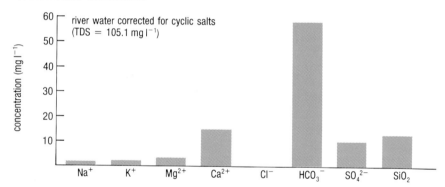

Figure 6.10 The average chemical composition of river water after 'correction' for cyclic salts (cf. Figure 6.8(b)).

QUESTION 6.9 When the composition of river water is corrected for cyclic salts, as in Figure 6.10, the proportions of the major ions differ from those in seawater by an even greater amount than shown in Figure 6.8. Which of the following relative cation and anion abundance patterns apply to river water (corrected for cyclic salts) and which to seawater?

(i) $Na^+(aq) > Mg^{2+}(aq) > Ca^{2+}(aq)$

(ii) $Ca^{2+}(aq) > Mg^{2+}(aq) > Na^+(aq)$

(iii) $HCO_3^-(aq) > SO_4^{2-}(aq) > Cl^-(aq)$

(iv) $Cl^-(aq) > SO_4^{2-}(aq) > HCO_3^-(aq)$

6.2.3 ORIGIN OF THE CHLORIDE

It is easy enough to see how the major ions sodium (Na^+), potassium (K^+), magnesium (Mg^{2+}) and calcium (Ca^{2+}) can be derived from weathering of rocks, because these elements have a high abundance in the Earth's crust (Table 3.2). By contrast, only a negligible proportion of the chloride (Cl^-) in river water comes from weathering, which is why we are justified in assuming all chloride in river water comes from cyclic salts (Figure 6.10).

So where does the chloride come from originally?

The answer lies in volcanism. Hydrogen chloride (HCl) is an important constituent of volcanic gases. Early in the Earth's history, volcanism was more widespread than it is now, because the Earth as a whole was hotter. Large quantities of this very soluble gas were emitted and quickly dissolved in the oceans. Chloride is classified as an **excess volatile** because its concentration in seawater cannot be accounted for by rock weathering.

6.2.4 THE SODIUM BALANCE

Are there other constituents in seawater for which a source other than weathering should be sought, and if so, how can we tell?

A means of testing for an additional source of an element in seawater is to make a *mass balance calculation*. The total amount of the element being added to the oceans by continental weathering is compared with the amount in seawater. If there is more of the element in seawater than can be accounted for by rock weathering, then there must be an additional source of that element. If the amount in seawater is the same or less, it is not necessary to seek a source additional to rock weathering, though such a source may well also exist.

One of the simplest approaches is to work out the **sodium balance**. We assume that there is no source other than rivers for sodium in seawater, and calculate the amount of continental crustal rock that has to be weathered to provide this sodium. We can then compare the results for other elements with that for sodium. The calculation involves a number of simplifying assumptions, but this does not matter because we wish merely to identify those elements in seawater for which there is likely to be an additional source.

The first step is to estimate how much crustal rock must be weathered to provide the sodium in 1 litre of seawater. Tables 3.1 and 6.1 give an approximate average of 11 g of sodium per litre of seawater, which is accurate enough for our simple first order sums. The average concentration of sodium in crustal rocks is 2.4% (Table 3.2), and we can assume that to be representative of continental crust. So, there are 2.4 g of sodium in 100 g of average continental crustal rock.

It is estimated that approximately three-quarters of the sodium in rocks is dissolved during weathering and eventually contributes to the seawater solution. The rest remains chemically combined in the minerals of detrital sediments (sands and clays).

QUESTION 6.10   75% of 2.4 is 1.8. So from every 100 g of weathered rock, about 1.8 g of sodium goes into solution. To the nearest 100 g, how many grams of rock must be weathered to provide the 11 g of sodium in 1 litre of seawater?

The next step is to see whether the amount of average crustal rock that provides the sodium in a litre of seawater can also provide the other dissolved constituents in that same litre. This has been done for selected elements in Table 6.3. The last column of Table 6.3 shows that the 'percentage in solution' for sodium is close to 75, which provided the basis of the calculation in Question 6.10.

Now examine the 'percentage in solution' figures for the four elements below sodium in Table 6.3. Do they suggest that continental weathering is an adequate source of supply for these elements in seawater?

According to Table 6.3, less than 10% of any of those elements in average crustal rock is required to go into solution on weathering, in order to account for their concentrations in seawater. Perhaps you can see why the approximations used in the sodium balance calculation do not matter greatly: you could increase the 'percentage in solution' figure for sodium to 100, or reduce it to 50 or even less, and it would not affect the conclusion we have just reached.

Table 6.3 Comparison of elements in seawater and continental crustal rock.

Element	In continental crust		In seawater (g l^{-1})	% in solution *
	%	g in 600 g rock		
Na	2.4	14.4	10.76	74.7
K	2.1	12.6	0.387	3.1
Ca	4.1	24.6	0.413	1.7
Mg	2.3	13.8	1.294	9.4
Sr	0.038	0.23	0.008	3.5
Se	5×10^{-6}	3×10^{-5}	$\sim 10^{-7}$	0.3
As	2×10^{-4}	1.2×10^{-3}	$\sim 10^{-6}$	0.08
Pb	1.25×10^{-3}	7.5×10^{-3}	$\sim 10^{-9}$	0.00001
Zn	7×10^{-3}	4.2×10^{-2}	$\sim 10^{-7}$	0.0002
Cu	5.5×10^{-3}	3.3×10^{-2}	$\sim 10^{-7}$	0.0003
Co	2.5×10^{-3}	1.5×10^{-2}	$\sim 10^{-9}$	0.000007
Cl	0.013	0.078	19.353	24 800
S	0.026	0.156	0.885	567
Br	0.00025	0.0015	0.067	4470
B	0.0003	0.0018	0.0046	256

$$* \text{ Percentage in solution} = 100 \times \frac{\text{g per litre seawater}}{\text{g per 600 g rock}}$$

The elements in the first group of Table 6.3 are major constituents in seawater. Elements in the second group are minor and trace constituents, and their 'percentage in solution' values are very small – well below 1. Indeed, at first sight it is surprising that for the first two groups in Table 6.3 the figures in the last column are mostly so small. They seem to imply that for many elements only a minute proportion is dissolved on weathering.

A more likely explanation is that such dissolved constituents are rapidly removed from seawater: the lower the 'percentage in solution', the more efficient the inorganic or biological removal processes are likely to be – and the shorter the residence time of a particular constituent in the ocean. This is the only way of reconciling the information for river water and seawater in Figures 6.8 and 6.10. It is quite obvious that seawater is not simply a more concentrated form of river water. If it were, then HCO_3^- rather than Cl^- would be the principal anion, and Ca^{2+} rather than Na^+ the principal cation. It follows that the residence times of calcium and carbon in the oceans are much shorter than those of sodium and chloride (see Section 6.2.5).

The last group in Table 6.3 are all anion-forming elements in seawater. Headed by chloride, these obviously represent the excess volatiles, those constituents whose concentrations in seawater cannot be accounted for by rock weathering alone (Section 6.2.3). Sulphur dioxide (SO_2), hydrogen bromide (HBr) and volatile boron (B) compounds are all known to be components of volcanic gases, along with CO_2, nitrogen, argon, hydrogen, and of course HCl and H_2O. Water in present-day volcanic gases is mainly recycled from the atmosphere and hydrosphere, much of it via subducted oceanic crust (see Figure 7.1); but most of the H_2O in volcanic gases probably originated from the planet's interior early in the Earth's history, as outlined at the start of Chapter 1.

More sophisticated mass balance calculations take into account factors such as the composition of volcanic gases and the composition and rate of deposition of marine sediments. These calculations suggest sources additional to weathering for some minor and trace constituents of seawater, such as selenium (Se), arsenic (As) and lead (Pb), even though they have very low 'percentage in solution' values in Table 6.3. Another interesting case is that of manganese (Mn): deep-sea sediments contain greater concentrations of manganese than continental rocks, and more manganese is being deposited than is supplied by weathering.

Can you suggest another source for manganese (and other elements) in the oceans, one that has only been discovered comparatively recently?

As you read at the start of Chapter 3, hydrothermal activity at ocean ridge crests and other sites of oceanic volcanism is known to supply some elements to seawater, including manganese.

6.2.5 CHEMICAL FLUXES AND RESIDENCE TIMES

Implicit in the arguments set forth in preceding Sections is the assumption that the oceans are chemically in a long-term **steady state**. This means that the rate of addition of dissolved constituents to seawater is balanced by their rate of removal, so that concentrations do not change significantly with time. (A useful analogy is with a bottling machine: empty bottles go in at one end, full bottles come out at the other, but the number on the conveyor belt is always the same, no matter how fast the machine is moving, and photographs of it taken at different times would all look alike.) There is evidence that a chemical steady state condition has characterized the oceans since early in the Earth's history, and that the composition of seawater has not varied significantly over the past several hundred million years.

The concept of residence time was introduced in Section 1.2, in relation to the hydrological cycle. How can it be applied to dissolved constituents in the oceans?

If the oceans are in a steady state, then the rates of supply and removal of dissolved constituents must be equal. The residence time of a dissolved constituent is given by:

$$\frac{\text{total mass dissolved in the oceans}}{\text{rate of supply (or removal)}}$$

and as rates of supply or removal are usually annual rates, residence times are normally given in years.

A traditional starting point for residence time calculations has been that rivers are the only source of supply of dissolved constituents. This basic assumption is still valid enough for most dissolved constituents to provide reasonable estimates of residence times, even though it is now known that hydrothermal solutions provide significant amounts of some elements.

The residence time of an element in seawater can therefore be estimated by dividing its mass in the oceans by its annual input from rivers. The annual flux of each element into the oceans via rivers can readily be calculated from the total annual inflow of river water to the oceans (e.g. Figure 1.3), multiplied by the average concentration of that element in river water (e.g. Figure 6.10). The total mass of each element in the oceans is also easily calculated from data for average concentrations in seawater and the total mass of water in the oceans (cf. Table 6.1). Oceanic residence times for several elements are presented in Table 6.4.

Table 6.4 River fluxes and residence times of some dissolved constituents in seawater.

Constituent	River flux* ($\times 10^8$ t yr^{-1})	Mass in ocean† ($\times 10^{14}$ t)	Residence time ($\times 10^6$ yr) (uncorrected)*	(corrected)
Na$^+$	2.05	144	70.2	210
K$^+$	0.75	5	6.7	10
Ca^{2+}	4.88	6	1.23	1.4
Mg^{2+}	1.33	19	14.3	22
Cl$^-$	2.54	261	103	(∞)
HCO$_3^-$	18.95	1.9	0.1	0.1
SO$_4^{2-}$	3.64	37	10.2	11
SiO$_2$	4.26	0.08	0.02	
Fe	0.01	0.000 001	0.000 1	
Zn	0.000 7	0.000 006	0.009	
Mn	0.000 04	0.000 000 4	0.001	

* These values are not corrected for cyclic salts.

† Amounts differ somewhat from those in Table 6.1 (see the accompanying text to that Table).

QUESTION 6.11 Major constituents in Table 6.4 have two residence times: on the left are values derived using the river flux uncorrected for cyclic salts (i.e. using concentrations in river water such as those in Figure 6.8); while on the right are values obtained after correcting the river flux for cyclic salts (i.e. using concentrations in river water such as those in Figure 6.10).

(a)   For how many constituents in Table 6.4 is the residence time significantly affected by this correction?

(b)   It has been estimated that some 2.5×10^8 t of calcium are added to the oceans annually by hydrothermal circulation. How would that affect the residence times in Table 6.4?

Residence times in Table 6.4 are approximate only. The averages used conceal wide variations and we have seen that the basic assumption is not wholly valid: rivers are not the sole source of dissolved constituents in seawater. The real residence time of chloride is neither infinite nor about

100 million years (Table 6.4), but somewhere in between. There is evidence that chloride is removed from seawater during hydrothermal circulation and, of course, it is also removed when **evaporite** salts are deposited from seawater evaporating in shallow enclosed sedimentary basins (cf. Figure 3.1). These losses are balanced by the continued emission of HCl from volcanoes, which adds chloride to the atmosphere–ocean system, and by weathering of uplifted oceanic crust and evaporites. Thus, rock weathering makes *some* contribution to all excess volatiles in seawater. Even for chloride, the concentration in river water cannot fall exactly to zero after correction for cyclic salts (cf. Figure 6.10). Another example is sulphate (SO_4^{2-}): its 'percentage in solution' (Table 6.3) is far too high to be accounted for by rock weathering alone. However, its concentration in river water changes very little when corrected for cyclic salts (compare Figures 6.8 and 6.10) and its residence time is hardly affected (Table 6.4). That is because sulphate is supplied to river water partly from weathering and decomposition of sulphate and sulphide minerals, and partly via the atmosphere, as outlined in Section 6.1.3 (Table 6.2).

Figure 6.11 shows that there is a broad correlation between residence time and concentration in seawater. Most of the major constituents (for which the 'percentage in solution' is high, Table 6.3) have long residence times and

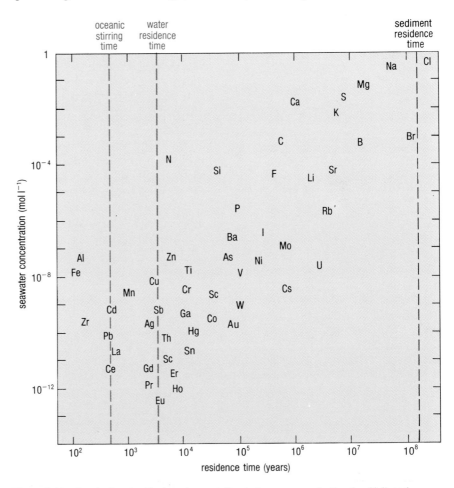

Figure 6.11 Graph showing the broad correlation between concentration ($mol\,l^{-1}$) and residence time for several elements in seawater. *Note that both scales are logarithmic.* For water residence time, see Question 1.3(b) and its answer; for oceanic stirring time and sediment residence time, see the following text.

remain in seawater for the order of 10^6–10^8 years, whereas many minor and trace constituents have short residence times and are removed from seawater in 10^3–10^4 years or less – and their 'percentage in solution' is low.

The oceanic stirring time for seawater (also called the turnover time, mixing time or exchange time) is shown as about 500 years in Figure 6.11. This represents the average time water spends in the deep oceans, before it returns to the surface. However, radiocarbon (carbon-14) dating of dissolved carbon species in samples of deep water from parts of the northern Pacific and Indian Oceans has yielded ages in excess of 1 000 years; while in other regions the exchange time between surface and deep oceans is closer to our average, e.g. about 300 years for the Atlantic as a whole and about 600 years for the Pacific as a whole.

For individual dissolved constituents, the residence time is the average length of time spent in the seawater solution. Dissolved constituents are added from rivers and other *sources*. They reside in solution for a time before being removed to the sediments and rocks of the sea-floor, which are the *sinks*. The reactions which remove dissolved constituents to the sinks are sometimes called reactions of **reverse weathering**, because they have the reverse effect to reactions which supply elements to seawater as a result of weathering on land (e.g. reactions 6.4 and 6.5). Removal mechanisms include inorganic precipitation and reactions between dissolved material and solid particles (both at the sea-bed and during diagenesis within sediments, Section 3.1.1), as well as biological processes. Marine organisms can concentrate minor and trace elements to very high levels in their soft tissues (see Section 6.3.4) and thus contribute to their removal from seawater; however, removal will be only temporary if the organism decomposes in the water column (as usually happens) rather than being preserved in sediments. The residence time for sediments in Figure 6.11 is longer than for most dissolved constituents, because it approximates to the lifetime of ocean basins. This is the length of time that a piece of sea-floor exists, on average, before being subducted at a destructive plate margin (see Figure 7.1), and is generally between 100 and 200 million years.

QUESTION 6.12 (a) What is the distinction between the oceanic stirring time and the residence time for water in the oceans?

(b) Are there any dissolved constituents whose residence times are too short to allow them to be mixed throughout the oceans?

The answer to Question 6.12(b) raises the further question of how dissolved constituents with short residence times can occur throughout the oceans if their only source is rivers. In fact, the major source of such constituents – especially iron and aluminium – is mineral particles in wind-blown (aeolian) dust carried from land far out to sea.

6.3 CHEMICAL AND BIOLOGICAL REACTIONS IN SEAWATER

The form or chemical **speciation** of dissolved constituents in seawater (see fourth column of Table 6.1) is very important in determining how they interact, and this in turn determines how long they remain in solution. Most of the dissolved constituents are in ionic form. The ions are kept apart because water has a high dielectric constant (Table 1.1), and each ion is surrounded by a sheath of water molecules called a **hydration sphere**,

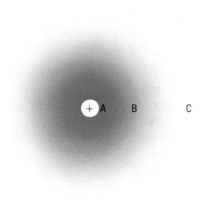

Figure 6.12   Pictorial representation of hydration of an ion. Water molecules in zone A are tightly held, in zone B less strongly bound, and in zone C hardly affected at all. There is probably a fairly well-defined boundary between zones A and B, but that between B and C is very diffuse.

which has a diffuse outer boundary (Figure 6.12). The size of the hydration sphere depends on the radius and charge of the ion, which determine the charge per unit area, or charge density. So we can make some simple generalizations.

1   *Anions* typically have lower charge densities than cations, because they are generally larger than the parent atom or molecule, having gained one or more electrons. *Cations* are generally smaller than the parent atom, because they have lost one or more electrons, and so cations have larger hydration spheres relative to their size than anions.

2   The greater the charge on an ion of given radius, the larger its hydration sphere relative to the size of the ion.

QUESTION 6.13   Which of the ions listed below is likely to have the largest hydration sphere relative to its size, and which the smallest?

Mg^{2+} radius = 66 pm (pm = picometre = 10^{-12} m)

Na^+ radius = 97 pm

Cl^- radius = 181 pm

6.3.1   INTERACTIONS BETWEEN DISSOLVED SPECIES

Cations and anions in solution experience electrostatic attraction and/or repulsion depending on their ionic charges. Such interactions are inversely proportional to the square of the distance separating the ions, and will be vanishingly small in very dilute solutions where the ions are widely separated. In solutions as saline as seawater, however, interactions between dissolved ions cannot be ignored. It is these interactions which determine the speciation of dissolved constituents, and their overall effect is to decrease the availability of ions for chemical reactions, whether inorganic or biological.

Figure 6.13 summarizes in diagrammatic form the three main types of interaction possible between ions in solution.

(a)   For ions of strongly ionic salts (strong electrolytes) such as sodium chloride, the only interaction is that of electrostatic attraction and/or repulsion between ions that otherwise behave as independent entities; the hydration spheres remain intact.

(b)   Some ions may form ion pairs, in which the hydration spheres of the constituent ions remain largely intact, and the result can be either a neutral or a charged species. It is uncommon for ion pairs to form between two monovalent ions; they are usually formed between two polyvalent ions or between one polyvalent and one monovalent ion.

(c)   Complex ion formation: although there is no clear division between the formation of complexes and ion pairs, there are two principal differences. First, the bonding in a complex is covalent, rather than electrovalent as in an ion pair; and secondly, when a complex is formed in solution the hydration spheres of the two or more entities making up the complex merge to form a joint hydration envelope.

In summary, if part of the total amount of a dissolved ion is interacting with other ions in solution, it will not be free to participate in chemical reactions. The extent of these interactions and their effect on the chemical reactivity of dissolved species depend on the nature of the ions involved. In general, for

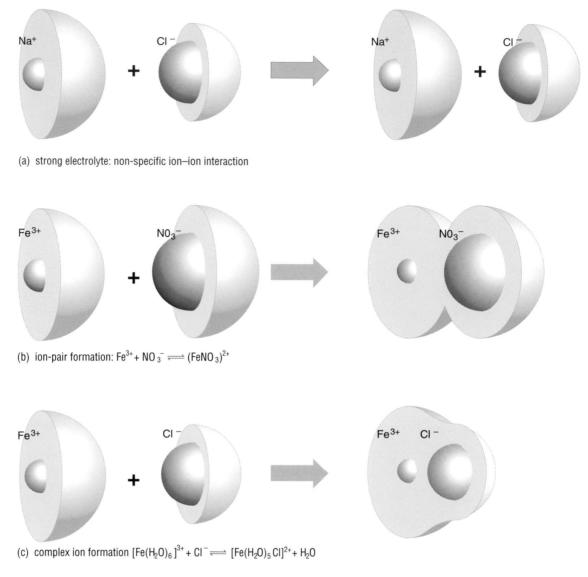

(a) strong electrolyte: non-specific ion–ion interaction

(b) ion-pair formation: $Fe^{3+} + NO_3^- \rightleftharpoons (FeNO_3)^{2+}$

(c) complex ion formation $[Fe(H_2O)_6]^{3+} + Cl^- \rightleftharpoons [Fe(H_2O)_5Cl]^{2+} + H_2O$

Figure 6.13   Diagrams illustrating different types of ionic interaction in seawater. Note that each ion is surrounded by a hydration sphere (cf. Figure 6.12). (a) General non-specific interaction. (b) Ion-pair formation. (c) Complex ion formation.

solutions of a concentration equivalent to seawater, the proportion of the total concentration of an ionic species that is free to react decreases as its charge increases.

The 'effective' concentration, therefore, is almost invariably less than the true concentration. For unassociated ions (e.g. Na^+), subject only to electrostatic interaction, this effective concentration (known as the *activity* in quantitative treatments of solution equilibria) may be in the region of 70% of the true concentration, whereas for some complex-forming multivalent ions (e.g. Al^{3+}) it may be as low as 5–10%.

There is a considerable degree of ion pairing between several of the eight most abundant dissolved ions in seawater: Na^+, K^+, Mg^{2+}, Ca^{2+}, Cl^-, HCO_3^-, CO_3^{2-} and SO_4^{2-}, which between them make up over 99% of the total. Concentration data and equilibrium constants for interactions between these four cations and four anions in solution have been used to calculate the results summarized in Table 6.5.

Table 6.5 Species distribution of major constituent ions in solution in seawater.

Ion	Concentration (mol l^{-1})	Free ion (%)	With SO_4^{2-} (%)	With HCO_3^- (%)	With CO_3^{2-} (%)
Calcium, Ca^{2+}	0.0104	91	8	1	0.2
Magnesium, Mg^{2+}	0.0540	87	11	1	0.3
Sodium, Na^+	0.4752	99	1.2	0.001	—
Potassium, K^+	0.0100	99	1	—	—

		Free ion (%)	With Ca^{2+} (%)	With Mg^{2+} (%)	With Na^+ (%)	With K^+ (%)
Sulphate, SO_4^{2-}	0.0284	54	3	21.5	21	0.5
Bicarbonate, HCO_3^-	0.00238	69	4	19	8	—
Carbonate, CO_3^{2-}	0.000269	9	7	67	17	—

QUESTION 6.14 (a) Which ion is missing from Table 6.5?

(b) Does this help to explain why a much smaller proportion of anions than cations appear to occur as free ions, according to Table 6.5?

(c) Which ion pair seems to be the most abundant, and in what context have you already encountered it?

6.3.2 THE CARBONATE SYSTEM, ALKALINITY AND CONTROL OF pH

The calcium carbonate used by many planktonic organisms to form their hard parts (Figure 6.3) redissolves when the organisms die and sink into deep water, releasing calcium and carbonate ions back into solution:

$$CaCO_3(s) \rightleftharpoons Ca^{2+}(aq) + CO_3^{2-}(aq) \tag{6.6}$$

Accordingly, should the ratio of dissolved calcium to total salinity (the $Ca^{2+} : S$ ratio) be greater in deep or in surface waters?

If calcium is extracted from surface waters and then returned to solution in deep waters, Ca^{2+} concentrations should be higher in deep than in surface waters. Calcium is a bio-intermediate constituent, but it is so abundant in seawater that its involvement in biological processes results in only small increases of the $Ca^{2+} : S$ ratio with depth (Section 3.1). These changes are small enough for calcium to be considered a conservative constituent of seawater for most purposes (Section 4.3.4), but they are very important in the context of the carbonate system.

Ocean surface waters are nearly everywhere supersaturated with respect to calcium carbonate, but in fact spontaneous inorganic precipitation of calcium carbonate occurs only infrequently. The reason for that lies in the inhibiting effect of Mg^{2+} ions: much of the carbonate in solution is in the form of $MgCO_3$ ion pairs (Table 6.5). It generally requires the intervention of marine organisms to precipitate the calcium carbonate. Calcareous skeletal material is made of either **calcite** or **aragonite**, which have the same chemical formula, $CaCO_3$, but different crystalline structure. The aragonite structure is thermodynamically less stable than that of calcite, so aragonite dissolves more readily than calcite.

Deep ocean waters, on the other hand, are everywhere *under*saturated with respect to calcium carbonate, partly because of the effect of pressure on the solubility of CO_2 (Section 6.1.3) and partly because the solubility of $CaCO_3$ itself also increases with pressure; so calcium carbonate dissolves. The depth at which significant dissolution of calcareous skeletal material begins (i.e. the depth where the water has become significantly undersaturated with respect to $CaCO_3$) is called the **lysocline**. The depth at which most or all of the $CaCO_3$ has dissolved is called the **carbonate compensation depth (CCD)** (not to be confused with the compensation depth defined for oxygen production/consumption, Section 6.1.3). Variations in depth of the lysocline are controlled by the chemistry of the water column (carbonate equilibria and pH, see below). Variations in the CCD are controlled partly by chemistry and partly by the rate of supply of calcareous material sinking from the surface.

In fact, most dissolution of calcium carbonate takes place actually *at the sea-bed itself*, because skeletal material usually sinks too fast for significant dissolution to occur *en route* to the bottom. The CCD can be thought of as a 'snowline' on sea-floor topography, whereas the lysocline can be located within the water column by reference to water chemistry, as noted above.

The CCD is thus the depth where the rate of dissolution of calcium carbonate skeletal material at the sea-bed equals (compensates for) the rate of supply of material sinking from the surface. No calcium carbonate should be preserved below this depth because, as we have seen, the solubility of calcium carbonate in seawater (i.e. the degree of undersaturation of seawater with respect to calcium carbonate) increases with depth below the lysocline. Therefore, the deeper the sea-bed, the more rapidly the calcium carbonate dissolves.

Other things being equal, the CCD lies deeper beneath areas of high biological production compared with areas of low production.

Why is that?

High biological production means large populations of organisms and a high rate of supply of calcareous skeletal material to deep water when the organisms die; whereas only a meagre supply of calcareous material sinks from regions of low biological production. Accordingly, for significant dissolution of a heavy 'snowfall' of calcium carbonate skeletal debris to occur, the sea-bed must be deeper than in an area where only a light 'drizzle' of calcareous material arrives from the surface. In short, the CCD tends to be depressed – i.e. the 'snowline' is deeper – beneath areas of high biological production compared with areas of low production.

The lysocline and CCD are shallower for aragonite than for calcite, and unless otherwise specified, the terms usually refer to calcite, because skeletal material is much more commonly formed of calcite than aragonite. Sediments below about 4 km depth seldom contain significant amounts of calcite, and it is rare to find aragonite remains in sediments below about 1–2 km. Both the lysocline and CCD are depth *zones* rather than precisely defined levels, because rates of dissolution depend also on factors such as turbulence in the water column and the nature of the skeletal debris (e.g. coccoliths are more delicate than foraminiferan tests; Figure 6.3(a) and (b)). In practice, the CCD is often defined as the depth where sediments contain less than about 10–20% of calcium carbonate.

At this point we must digress briefly to record that the behaviour of *silica* (SiO_2) is different from that of calcium carbonate in the water column. The solubility of silica decreases as temperature falls, so silica should become less soluble with depth; but this effect is offset by the increase of solubility with pressure (depth). The net result is that although the skeletal remains of organisms such as diatoms and radiolarians (Figure 6.2(a) and (b)) progressively dissolve as they sink through the water column (cf. Figure 6.1(c)), the rate of dissolution is relatively slow and much siliceous debris reaches the deep ocean floor; i.e. there is no compensation depth for silica.

We now need to look at the behaviour of carbon in more detail. The $C:S$ ratio is a good deal more variable than the $Ca^{2+}:S$ ratio (Sections 3.1 and 4.3.4), but in general it changes in the same direction: it is lower in surface than in deep waters.

Why might that be?

The ratio is influenced partly by the formation and subsequent dissolution of calcium carbonate skeletal material; but chiefly by the formation of soft organic tissue during primary production in surface waters and its subsequent decomposition as it sinks from the surface.

As you know from reaction 6.2 (Section 6.1.3), carbon occurs as several species in solution: dissolved CO_2 gas, $H_2CO_3(aq)$, $HCO_3^-(aq)$, and $CO_3^{2-}(aq)$. Together these make up the **total dissolved inorganic carbon**, expressed as ΣCO_2 (where Σ is capital sigma, and denotes 'sum of'). Some dissolved carbon is also combined in organic molecules, but this carbon does not take part in reactions of the carbonate system and its amount is very small in relation to total dissolved *inorganic* carbon, ΣCO_2. For simplicity, we shall generally refer to ΣCO_2 simply as total dissolved carbon. Figure 6.14(a) shows typical profiles for total dissolved carbon, and illustrates how carbon is the least conservative of the major dissolved constituents with obvious bio-intermediate character.

You read in Section 6.1.3 that CO_2 *as dissolved gas* is present in very small amounts in seawater. This concentration of CO_2 *as dissolved gas* increases only slightly with depth. So, when you read or hear about CO_2 concentrations in seawater, you should realize that this is nearly always simply a shorthand way of describing the concentration of total dissolved carbon. It is *not* describing the concentration of dissolved CO_2 gas, even though CO_2 happens to be the most convenient form in which to analyse total dissolved carbon, as we shall see. To be sure, the profile of increasing ΣCO_2 with depth in Figure 6.14(a) is largely the result of the production of CO_2 in respiration and decomposition of organic matter (reaction 6.3 going to the left). But the CO_2 produced is *not* liberated as bubbles of gas, because as soon as it forms it combines with water in reaction 6.2. By far the most important components in that reaction are HCO_3^- and CO_3^{2-}, and it is to the relationships between these two ions that we now turn.

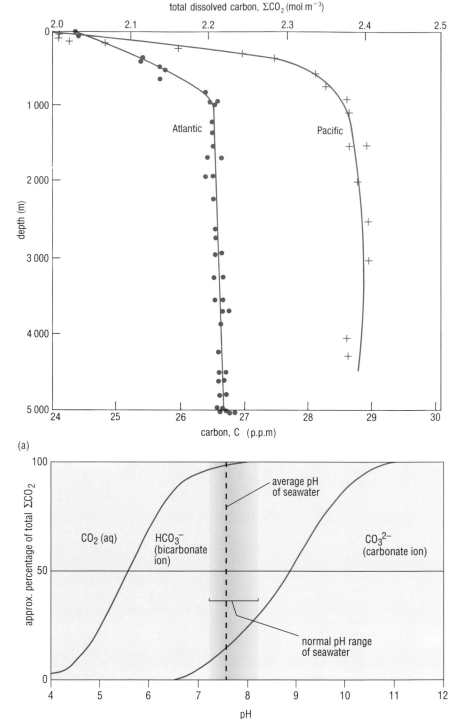

Figure 6.14 (a) Variation with depth of total dissolved inorganic carbon expressed as ΣCO_2 and given in mol m^{-3}, as well as p.p.m. of carbon, in the Atlantic at 36° N, 68° W (dots), and the Pacific at 28° N, 122° W (crosses). *Note*: to convert mol l^{-1} to mol m^{-3}, simply multiply the concentration by 10^3, e.g. 2×10^{-3} mol l^{-1} becomes 2 mol m^{-3}. Note also that the horizontal axis does not start at zero, and that the concentration increase from surface to deep water is only about 10–20%.

(b) Generalized diagram showing approximately how relative proportions of three principal components in the aqueous carbonate system vary with pH in natural waters. The pH of seawater averages about 7.7 and can *range* from about 7.2 to 8.2. Positions of the curved lines vary with temperature, salinity and pressure.

* Regardless of pH, the concentration of $H_2CO_3^-$ in aqueous solution is always very low.

Alkalinity

The equilibria in reaction 6.2 are such that at the pH of seawater, which is generally close to 8 (Section 6.1.3), the carbonate system in solution is dominated by HCO_3^-, along with some CO_3^{2-}, but no $H_2CO_3^*$ (and very little dissolved CO_2 *gas*, as we have seen). Figure 6.14(b) summarizes relationships between these three components of reaction 6.2 and shows that the bicarbonate : carbonate ratio (HCO_3^- : CO_3^{2-}) increases as pH falls (more acid) and decreases as pH rises (less acid).

According to Figure 6.14(b), approximately what is the percentage of carbonic acid, H_2CO_3, at (i) the lower end and (ii) the upper end of the seawater pH range shown?

It is (i) about 5% at the lower end of the pH range (more acid, pH 7.2) and (ii) zero at the upper end (less acid, pH 8.2).

It is difficult to measure the relative proportions of bicarbonate and carbonate ions in solution directly, but their combined concentrations can easily be determined by laboratory titration with acid. This simply involves progressively adding acid to seawater samples until the negative charges are neutralized (the end-point of the titration) and all the bicarbonate and carbonate ions have been converted to water and CO_2, which escapes from the solution as gas:

$$HCO_3^-(aq) + H^+(aq) \rightarrow H_2O + CO_2 \text{ (gas)} \tag{6.7}$$

$$CO_3^{2-}(aq) + 2H^+(aq) \rightarrow H_2O + CO_2 \text{ (gas)} \tag{6.8}$$

Why does it require two moles of hydrogen ion to neutralize the charge on one mole of carbonate ion (reaction 6.8)?

Because the carbonate ion carries two negative charges. Only one mole of hydrogen ion is required to neutralize a mole of the singly charged bicarbonate ion (reaction 6.7).

The amount of acid used in the titration is a measure of the total amount of hydrogen ions required to neutralize the negative charges on the carbonate and bicarbonate ions in solution (equations 6.7 and 6.8). One mole of hydrogen ion is required to neutralize one mole of negative charge.

The combined molar concentration of carbonate and bicarbonate ions, expressed in 'charge-equivalent' terms (and determined by titration with acid as outlined above), is known as the **alkalinity**, A, of seawater. Usage of the term alkalinity in this context differs from common chemical usage in that it is not a measure of how 'alkaline' – as opposed to 'acid' – the water is. Relationships between alkalinity and pH are developed below.

Important: Square brackets around chemical formulae are the conventional symbol for concentration in solution. So, for example, $[CO_3^{2-}]$ simply means the concentration of carbonate ion in solution. In this discussion, we shall use concentrations expressed in $mol\,m^{-3}$ ($= mol\,l^{-1} \times 10^3$).

Note: It is not essential to follow all of the discussion below. If your experience of chemistry is limited, you may wish to continue with the subsection headed 'ΣCO_2, alkalinity and pH', bearing in mind that you should be able to use equations 6.10, 6.11 and 6.15 in calculations involving alkalinity.

We have just defined alkalinity as the total negative charge carried by the carbonate and bicarbonate ions in solution, expressed in terms of molar concentrations. From this definition of alkalinity, it can be expressed as:

$$A = [HCO_3^-] + 2[CO_3^{2-}], \tag{6.9}$$

and alkalinity has units of molar concentration ($mol\,m^{-3}$ in this discussion).

Why is the concentration term for carbonate ion doubled in equation 6.9?

It is doubled because (from equation 6.8) two hydrogen ions are required to neutralize the two negative charges on the carbonate ion.

If the gaseous CO_2 produced in the titration (equations 6.7 and 6.8) is collected and measured, its amount provides a value for the concentration of total dissolved inorganic carbon in the seawater sample, which we can express as (cf. Figure 6.14 (b)):

$$[\Sigma CO_2] = [HCO_3^-] + [CO_3^{2-}] \tag{6.10}$$

Why is the concentration term for carbonate ion *not* doubled in equation 6.10?

It is not doubled because one mole of carbonate ion produces one mole of molecular CO_2, even though it requires two moles of hydrogen ion to do so (equation 6.8).

If we subtract equation 6.10 from equation 6.9, we get:

$$A - [\Sigma CO_2] = [CO_3^{2-}] \tag{6.11}$$

The outcome of all this is to show that by measuring alkalinity and total dissolved carbon, equation 6.11 can be used to determine the carbonate ion concentration; and substitution of that value into equation 6.9 or 6.10 gives the concentration of bicarbonate ion. The bicarbonate : carbonate ratio can then be determined, and it is this ratio which provides the main control on the pH of seawater.

The control of pH

pH is a measure of the concentration of hydrogen ions in a solution (see also the Appendix):

$$pH = -\log_{10}[H^+] \tag{6.12}$$

In seawater, pH is mostly in the range 7.7 ± 0.2 (Figure 6.14(b)) and variations of pH are controlled chiefly by a component of reaction 6.2:

$$HCO_3^-(aq) \rightleftharpoons H^+(aq) + CO_3^{2-}(aq) \tag{6.13}$$

The reaction is very rapid and seawater can be assumed to have an equilibrium mixture of the three ions. When reaction 6.13 is at equilibrium, we can write:

$$K = \frac{[H^+][CO_3^{2-}]}{[HCO_3^-]} \tag{6.14}$$

where K is the equilibrium constant.

If we re-arrange equation 6.14:

$$[H^+] = K \frac{[HCO_3^-]}{[CO_3^{2-}]} \tag{6.15}$$

Equation 6.15 shows that the ratio of the concentration of HCO_3^- ions and CO_3^{2-} ions must control the hydrogen ion concentration and hence pH: as the ratio increases, so does $[H^+]$, and pH will decrease.

If you need to refresh your memory on how to work out pH from $[H^+]$ and *vice versa*, consult the Appendix.

At this point, an obvious question presents itself: why not measure pH directly? The reason is that seawater is, chemically speaking, a concentrated solution (Section 6.3.1) and reliable values of pH are not easily obtained by direct measurement.

The control of alkalinity

To understand how alkalinity can vary in the oceans, we need to use an alternative definition: Alkalinity is the net molar concentration, in 'charge-equivalents', of the cations of strong bases in solution *in excess* of the net molar concentration, in 'charge-equivalents', of the anions of strong acids in solution, i.e.:

$$A = [\text{strong base cations}] - [\text{strong acid anions}]$$

where square brackets indicate molar concentrations in 'charge-equivalents'.

So far as seawater is concerned, that means:

$$A = [Na^+] + [K^+] + 2[Mg^{2+}] + 2[Ca^{2+}]) - ([Cl^-] + 2[SO_4^{2-}] + [Br^-] \qquad (6.16)$$

These are the dominant ionic constituents of seawater, other than bicarbonate (and carbonate), cf. Table 3.1. Note that, as before, double weighting is given to doubly charged ions, to express their concentrations in 'charge-equivalent' terms (cf. discussion of equation 6.9 above); remember also that seawater is electrically neutral (cf. Question 3.1), i.e. total positive charge = total negative charge.

If you were to work out the totals in equation 6.16, you would find that the value for A lies close to $2 \, \text{mol m}^{-3}$. In short:

$$A = [HCO_3^-] + 2[CO_3^{2-}] \qquad (6.9)$$
$$\approx 2 \, \text{mol m}^{-3} \text{ on average, throughout the oceans.}$$

If all of the dominant ionic constituents, other than bicarbonate and carbonate, behaved conservatively, would the alkalinity of seawater change?

No. It would remain constant. However, as summarized at the start of this Section, we know that Ca^{2+} is a bio-intermediate constituent (Section 6.1.2) and that the ratio $[Ca^{2+}]:S$ is slightly greater in deep than in surface water (Section 3.1).

What does that suggest about alkalinity in deep water relative to that in surface water?

Alkalinity must be greater in deep than in surface water because the Ca^{2+} extracted by marine organisms to form calcium carbonate tests in surface waters is returned to solution when the skeletal material re-dissolves in deep water. In other words, the total concentration of 'strong base cations' relative to 'strong acid anions' is greater in deep than in surface waters.

ΣCO_2, alkalinity and pH

A number of important conclusions follow from the foregoing discussions:

1   Total dissolved inorganic carbon (ΣCO_2) is relatively depleted in surface waters and relatively enriched in the deep ocean (Figure 6.14(a)), chiefly because of the formation of organic matter (soft tissue) by planktonic marine organisms and its subsequent decomposition; *and partly also* because of the formation and subsequent dissolution of calcium carbonate skeletal material.

2   Alkalinity, A, is lower in surface waters than in the deep ocean, *only because* of the formation and subsequent dissolution of calcium carbonate skeletal material – and it is important to recognize that by no means all plankton form calcium carbonate skeletons or tests.

3   In short, alkalinity is controlled *only* by the formation and dissolution of calcium carbonate, whereas ΣCO_2 is controlled *both* by formation and decomposition of organic matter (soft tissue) *and* by formation and dissolution of calcium carbonate.

4   Because of equations 6.9 and 6.10, *A* must always be slightly greater than ΣCO_2 for any particular sample of seawater. However, because not all planktonic marine organisms secrete calcium carbonate skeletons, ΣCO_2 increases with depth by a greater amount than *A* does. The effect of this on the way pH changes with depth is illustrated in Question 6.15.

QUESTION 6.15   (a)   The alkalinity of a sample of surface water is 2.35 mol m⁻³ and its [ΣCO_2] is 2.0 mol m⁻³; the same quantities for a sample of deep water from the same location are 2.55 mol m⁻³ and 2.4 mol m⁻³. Use equations 6.11, 6.10 and 6.15 (in that order) to work out the pH for these two samples. Use a value of 1.0×10^{-9} for the equilibrium constant *K,* and note that [H⁺] in equation 6.15 will be expressed in mol l⁻¹ (concentrations are given in mol m⁻³ but the units will cancel out in equation 6.15).

(b)   If there were no planktonic organisms to secrete calcium carbonate skeletal material in the open oceans (as is believed to have been the case in the geological past), then alkalinity would be effectively constant there. But ΣCO_2 would still increase with depth. With the help of equations 6.11, 6.10 and 6.15, explain why deep water would still be more acid than surface water.

In summary, the greater the [ΣCO_2], the smaller the value of ($A - [\Sigma CO_2]$), the greater the value of [HCO_3^-]/[CO_3^{2-}], the higher the value of [H⁺], the lower the pH, and the more acid the water.

The corollary is that where [ΣCO_2] is low, pH is high. For example, one of the few places where inorganic precipitation of calcium carbonate occurs is on the Bahamas Banks, where the sea is shallow and warm and the salinity is high (greater than 37). The warmer and more saline the water, the lower the solubility of gases, including carbon dioxide. In these conditions, the term ($A - [\Sigma CO_2]$) in equation 6.11 will be large.

What will that mean for the value of [CO_3^{2-}]?

The concentration of carbonate ions will also be large, which is consistent with relationships summarized in Figure 6.14(b). [CO_3^{2-}] can rise sufficiently for the water to be so supersaturated with respect to $CaCO_3$ that the inhibiting effect of the $MgCO_3$ ion pair is overcome, and small crystals of calcium carbonate (in the form of aragonite) are precipitated. Equations 6.9 to 6.15 thus help to explain what might seem at first sight a paradox: other things being equal, where [ΣCO_2] is high, calcium carbonate is more likely to dissolve; and where [ΣCO_2] is low, calcium carbonate is more likely to precipitate.

The relationships outlined above have been simplified in order to establish some basic principles of seawater chemistry. As usual, reality is not quite so simple. For example:

1   We have ignored the other forms of dissolved inorganic carbon as being quantitatively insignificant, but they cannot be neglected when accurate measurement and calculation are required.

2   The alkalinity defined above should (strictly speaking) be called *carbonate alkalinity*, because other ions, notably boron species such as

$H_2BO_3^-$, contribute to the total alkalinity of seawater, and it is in fact the total alkalinity that is measured by titration.

3   The equilibrium constants for the component reactions of the carbonate system (e.g. equation 6.15) are not strictly 'constant', but change with temperature and pressure.

However, none of these complications affects the basic principles. The concepts introduced in this Section are not easy, but they are important because of the central role of carbon in seawater chemistry and their relevance to the global CO_2 problem, discussed in Chapter 7. Before moving on, however, we must repeat the point made earlier. The alkalinity of natural waters is *not* a measure of how 'alkaline' – as opposed to 'acid' – those waters are. Usage of the term alkalinity in marine chemistry differs from common chemical usage, where only water of pH greater than 7 is considered to be alkaline. However, it *is* consistent with such usage in that when $(A - [\Sigma CO_2])$ is high, so is pH (less acid), and when $(A - [\Sigma CO_2])$ is low, so is pH (more acid).

Tabular and diagrammatic summaries of the relationship between alkalinity and ΣCO_2 are provided in Table 6.6 and Figure 6.15.

Figure 6.15 Plot of concentration of total dissolved inorganic carbon [ΣCO_2] against alkalinity, *A*. Concentrations in mol m^{-3}, i.e. (mol l^{-1}) × 10^3. Symbols ΔA and $\Delta\Sigma CO_2$ denote changes in alkalinity and total dissolved carbon respectively. For explanation of arrows and ovals, see Notes to Figure. See also Table 6.6.

Notes to Figure 6.15
• Horizontal green arrows (upper right) show that photosynthesis and respiration change *only* ΣCO_2, they do not change *A* (i.e. $\Delta A = 0$).
• Horizontal grey arrows (lower right) show that gain and loss of atmospheric CO_2 by gas exchange across the air–sea interface change only ΣCO_2, they do not affect *A* (i.e. $\Delta A = 0$).
• Inclined yellow–green arrows (upper left) show that formation and dissolution of $CaCO_3$ change *both* ΣCO_2 *and A*. In this case, $\Delta A = 2\Delta CO_2$, because for every mole of $CaCO_3$ formed or dissolved, ΣCO_2 is changed by 1 mol of CO_3^{2-}, whereas *A* is changed by 2 mol of CO_3^{2-} (i.e. because the concentration term [CO_3^{2-}] is doubled in equation 6.9 but not in equation 6.10).
• Alkalinity is changed *only* by formation and dissolution of $CaCO_3$.
• Blue ovals show relationship between ΣCO_2 and *A* for water from various parts of the ocean; see also Table 6.6.
• The difference between Warm Surface and Cold Surface fields results *only* from photosynthesis/respiration and gas exchange across the air–sea interface.
• The sequence from Cold Surface to Deep Pacific results from a combination of respiration (consumption of organic matter) and dissolution of $CaCO_3$.
• Both ΣCO_2 and *A* increase as amounts of organic matter consumed and of $CaCO_3$ dissolved increase. These amounts are greatest in Deep Pacific waters.
• $(A - [\Sigma CO_2])$ decreases progressively from Warm Surface to Deep Pacific, so pH decreases progressively, from 8.2 to 7.6.

Table 6.6 Carbonate chemistry of various seawater types (see also Figure 6.15).

Water type	[ΣCO_2] (mol m^{-3})	Alkalinity A [HCO_3^-] + 2[CO_3^{2-}] (mol m^{-3})	Bicarbonate ion [HCO_3^-] (mol m^{-3})	Carbonate ion [CO_3^{2-}] (mol m^{-3})
Warm surface	2.00	2.35	1.65	0.35
Cold surface	2.15	2.35	1.95	0.20
Deep Atlantic	2.25	2.40	2.10	0.15
Deep Indian	2.35	2.45	2.25	0.10
Deep Pacific	2.45	2.55	2.35	0.10

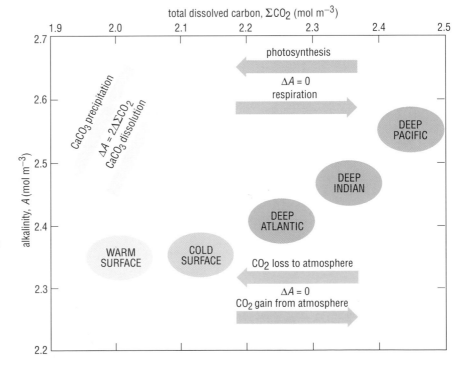

6.3.3 NON-BIOLOGICAL CONTROLS ON MINOR AND TRACE ELEMENT CONCENTRATIONS

Seawater is undersaturated with respect to the great majority of both major and minor constituents. As you read in Section 3.1.2, salts are not precipitated from seawater unless concentrations are greatly increased by evaporation. This is consistent with the long residence times and relatively high 'percentage in solution' values for the major constituents (Tables 6.3 and 6.4). However, minor and trace elements have *low* 'percentage in solution' values and short residence times (Table 6.3 and Figure 6.11) which means that they are removed quickly from the seawater solution – they move rapidly from source to sink. But how? Simple chemical precipitation cannot be the answer, because seawater is undersaturated in these elements. The following three non-biological removal mechanisms apply to the trace metals, which in fact make up the majority of minor and trace elements in seawater. The mechanisms should not be considered as being mutually exclusive.

1   *Mutual attraction* between the charges on ions in solution and small residual surface charges on suspended particles results in **adsorption** of metal ions (or ion pairs or complexes, cf. Figure 6.13) onto particles of both organic origin (detritus, bacteria) and inorganic origin (clay minerals, hydroxides). Elements adsorbed onto small particles are removed from the water column as large particles 'capture' the small ones and carry them downwards – a process called **scavenging** (many oceanographers use this term to encompass both parts of the process: adsorption and capture).

Particle adsorption and scavenging of dissolved ions is the most important non-biological mechanism for removing trace metals from seawater. Elements for which it is the principal removal process are characterized by concentration profiles showing a decrease in concentration with depth, as in Figure 6.16 – i.e. approximately a mirror image of nutrient-type profiles (Figure 6.1).

In the open ocean, bacteria dominate the seston to the extent that their combined total surface area exceeds that of inorganic particulate matter by about an order of magnitude – so bacteria are the dominant scavengers of trace metals. The process is purely physical, however: the bacteria are simply passive agents that provide a suitable surface for adsorption. No biological use is made of the scavenged elements.

Just how effective particle scavenging is as a mechanism for rapidly removing dissolved constituents from seawater was dramatically illustrated in the aftermath of the Chernobyl nuclear accident of April 1986. Sediment traps in the Mediterranean, North Sea and Black Sea recorded large increases in radionuclides at depths of 200 m or more, within days of the arrival of the 'radioactive cloud' overhead. The research showed that the radionuclides were rapidly adsorbed on to particles which were in turn ingested by zooplankton and aggregated into faecal pellets. These sank at rates of several tens of metres per day (cf. Question 6.2), taking the radionuclides with them – the radionuclides evidently passed rapidly through the digestive tracts of the animals. The radioactivity was removed from the surface to the sea-floor in a matter of weeks, or a few months at most. The radionuclides included isotopes of ruthenium (Ru), caesium (Cs) and cerium (Ce), the majority with half-lives of weeks, so they would probably not present a significant danger to the bottom-living animals (benthos). The exception is ^{137}Cs, with a half-life of 30 years, which could have longer-term effects where concentrations were high.

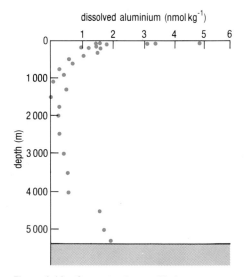

Figure 6.16   Concentration profile for dissolved aluminium in the central North Pacific (28°15' N, 155°07' W). The increase in concentration at the bottom of the profile may be due to re-solution in deep water and/or to diffusion from water in sea-bed sediments (nmol = nanomol = 10^{-9} mol, and 1 nmol l^{-1} $\approx$ 1 nmol kg^{-1}).

dissolved aluminium (nmol kg^{-1})

2   **Oxidation–reduction equilibria**: The oxidation state of elements with more than one valency can greatly affect their solubilities. The oxidation–reduction balance (or **redox** status) of natural waters can thus control the concentrations of such elements, which include several of the trace metals in seawater. For example, the oxidized form of iron, the trivalent form, called ferric iron or iron(III) and represented in solution by Fe^{3+}(aq) ions, is much less soluble than the reduced form, divalent iron, called ferrous iron or iron(II) and represented in solution by Fe^{2+}(aq) ions. That is because the concentration of Fe^{3+}(aq) ions in solution is limited by the very low solubility of ferric (iron III) hydroxide, $Fe(OH)_3$, compared with that of its iron(II) counterpart, ferrous hydroxide, $Fe(OH)_2$.

So in water sufficiently oxidizing for iron(III) to be dominant the total amount of dissolved iron will be very small because most of it will exist as Fe^{3+} in solid phases such as colloidal $Fe(OH)_3$, or the mineral goethite, $FeOOH$. Under more reducing conditions, the dominant valency state for iron will be $+2$, and much higher concentrations of soluble iron (as Fe^{2+}) are to be expected. Two other examples are cobalt and manganese, which are present in seawater as Co^{2+}(aq) and Mn^{2+}(aq) respectively, but are readily oxidized to less soluble Co^{3+}(aq) and Mn^{4+}(aq) and precipitated as hydroxides or hydrated oxides.

QUESTION 6.16   (a)   Is seawater normally an oxidizing or reducing medium?

(b)   Is the natural form of dissolved iron in normal seawater the more or the less soluble form?

(c)   Uranium is more soluble in its oxidized U^{6+}(aq) form than in its reduced U^{4+}(aq) form. Is the concentration of dissolved uranium likely to be greater or less in normal (oxic) seawater than in anoxic seawater?

3   *Co-precipitation* is another removal mechanism for trace metals, those at lower concentrations following more abundant constituents into precipitated phases, e.g. cobalt in the iron mineral goethite to give $(Fe,Co)OOH$, or lead in manganese oxide to give $(Mn,Pb)O_2$.

6.3.4   BIOLOGICAL CONTROLS ON MINOR AND TRACE ELEMENT CONCENTRATIONS

We have seen that biological activity has rather little effect on the concentrations of major constituents in seawater – even the important carbonate reactions described in Section 6.3.2 are associated with only small changes in total concentration. For minor and trace elements, however, it is a different story.

The ability of seaweeds, especially *Laminaria*, to concentrate iodine, as well as sodium and potassium (but only in very small amounts relative to the total concentrations of these elements), has led in the past to harvesting of the seaweed and extraction of these elements. (At present, seaweeds are collected on an industrial scale as a source of alginates which are used, amongst other things, as gelling and emulsifying agents in the food processing industry. This has nothing to do with their trace element content.)

Shellfish have long been known to concentrate trace metals, with enrichment factors of many thousands (where enrichment factor is defined as weight of element per unit weight of organism / weight of element per unit weight of seawater). It follows that where industrial effluents discharged at coastal sites contain metal concentrations above those of normal seawater, concentrations in shellfish may be correspondingly higher too, and this could make them toxic to organisms at higher levels in the food chain – including humans.

Plankton usually concentrate trace elements more strongly than do organisms further along the food chains. The following enrichment mechanisms have been suggested to occur either in isolation or collectively:

1   Ingestion of particulate suspended matter such as clays or organic particles which have scavenged minor or trace elements from seawater (Section 6.3.3). This will be most significant for filter-feeding organisms.

2   Ingestion of elements already concentrated in food material: plankton concentrate the trace elements and then species higher in the food chain eat the plankton. Such progressive concentration has so far only been shown to occur for mercury and manufactured organic compounds such as DDT and PCBs (polychlorobiphenyls).

3   Complexing of metals with organic molecules. At the mucous surfaces of digestive glands or gills of many organisms there are large molecules such as glycoproteins which can form complexes with metal ions.

4   Incorporation of metal ions into physiologically important systems, e.g. cobalt in vitamin B_{12}; copper in haemocyanin, the blood pigment of molluscs and crustaceans (cf. iron in haemoglobin); vanadium or niobium in the blood pigment of certain ascidians (sea-squirts); and titanium in some sponges. Accumulation of polonium (a radioactive decay product of uranium-238) in prawns of the genus *Gennadas*, can result in their receiving α-radiation doses that are twice the lethal limit for humans.

Biological processes thus clearly influence the trace element composition of seawater, and a number of trace elements exhibit biolimiting or bio-intermediate behaviour. Two examples are shown in Figure 6.17.

QUESTION 6.17   (a)   In most parts of the ocean, copper is identified as exhibiting bio-intermediate behaviour. How does Figure 6.17(a) suggest that copper could be a biolimiting micro-nutrient? (The term micro-nutrient is used where the element concerned is a minor or trace constituent in seawater.)

(b)   How is it possible to infer from Figure 6.17(b) that nickel distribution in ocean waters may be biologically controlled?

The redox state of trace elements (Section 6.3.3) affects their behaviour in marine biological systems. For example, selenium is more readily available to marine organisms as Se(VI) than as Se(IV) (represented by SeO_4^{2-}(aq) and SeO_3^{2-}(aq) ions respectively); whereas arsenic is generally a good deal more toxic as As(III) than as As(V) (AsO_3^{3-}(aq) and AsO_4^{3-}(aq) respectively).

Redox equilibria are of course not the only factor controlling speciation of elements in seawater and hence their availability to marine organisms. Thus, until the 1960s, mercury was thought to take part in marine biological processes only in inorganic forms, but in 1963 (the Minamata Bay tragedy in Japan) it was found that numerous cases of mercury poisoning, some fatal, were due to eating locally caught fish that had accumulated organic mercury compounds from industrial effluents. It is now known that mercury is more readily taken up by marine organisms when in the form of organic complexes (especially methyl mercury), than in its simple ionic forms (Hg^+, Hg^{2+}). The same is true of lead: marine organisms 'prefer' organic lead complexes (especially alkyl lead) to the simple ionic forms (Pb^{2+}, Pb^{4+}).

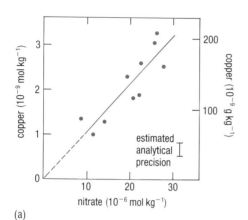

(a)

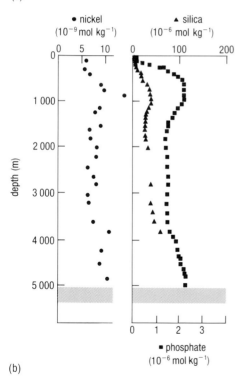

(b)

Figure 6.17   (a) Plot of nitrate versus copper concentrations in water samples from the Antarctic Ocean, showing a clear covariance. (b) Profiles for nickel, phosphate and silica in the tropical Atlantic showing that nickel distribution generally follows that of phosphate and silica. Note that the nickel concentrations (dots) are in 10^{-9} mol kg^{-1}; phosphate (squares) and silica (triangles) are in 10^{-6} mol kg^{-1}.

6.3.5 BIOLOGICAL ACTIVITY AS A SINK FOR TRACE ELEMENTS

Figure 6.18 summarizes the fate of the carbon fixed in organic matter by photosynthetic primary production in the photic zone. About 90% of the organic matter forming the soft parts of the phytoplankton is recycled above the thermocline because of consumption by animals and bacterial decomposition of detritus and excretion products. Most of the remaining 10% is recycled as it sinks towards the sea-floor. Only a small fraction reaches the bottom, and most of that is consumed or decomposed by the deep-sea benthos (bottom-living organisms). Very little organic matter is preserved in the sediments.

Recycling of organic matter in the water column involves the re-conversion (re-mineralization) of carbon and nutrients in organic compounds back into inorganic species in solution (bicarbonate, nitrate, phosphate), where they are again available for biological (primary) production.

The minor and trace elements taken up by marine organisms will in general share the same fate as the organic carbon, because they are concentrated mainly in the soft tissues. However, the concentration of these elements in the sinking organic detritus probably increases progressively with depth. This can happen in two ways. First, the detritus becomes more *refractory*, as organisms feeding on it make use of the nutritious components and reject the rest. Secondly, the particles adsorb and scavenge trace elements from solution as they sink through the water column (Section 6.3.3). Minor and trace elements concentrated in skeletal material will partly be returned to the water column by re-solution, and partly preserved in sediments.

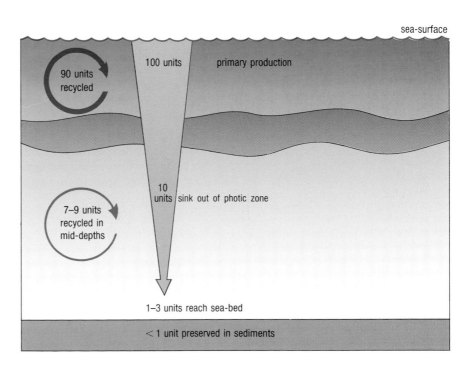

Figure 6.18 Sketch (not to scale) to illustrate the progressive decrease with depth of the organic carbon fixed by primary production in the photic zone.

The activities of organisms may thus provide a sink for minor and trace elements that is as important as the 'inorganic' sinks outlined in Section 6.3.3. It can be a great deal more important in coastal regions of *upwelling* where nutrients are abundant and there is high productivity (Section 6.1.2). In such conditions, the supply of dissolved oxygen may be insufficient to oxidize fully the enhanced 'rain' of organic particles, and metal-containing organic compounds may not be decomposed. Their burial in the sediment becomes a permanent sink for the trace elements they contain. Trace element enrichment in organic sediments (e.g. black shales) indicates that this process can be of considerable local importance.

Where does all this discussion of sinks for minor and trace elements leave the concept of the steady-state ocean?

Bear in mind the analogy of the bottling factory and its conveyor belts. Constituents removed from the seawater solution and preserved in sediments are replenished from various sources: river input, atmospheric fall-out and hydrothermal activity. In fact, there is continuous exchange of elements between seawater and the ocean floor, as well as between atmosphere and ocean. Such exchanges at ocean boundaries have been going on ever since the Earth acquired its atmosphere and oceans.

6.4 SUMMARY OF CHAPTER 6

1   There are eleven major dissolved constituents of seawater, with concentrations greater than 1 part per million by weight (1 in 10^6). The remainder are minor and trace constituents, the boundary between the two being about 1 part per billion by weight (1 in 10^9). Most of the known chemical elements have been found in the seawater solution; it is likely that all are present and will eventually be detected. The boundary between what constitutes dissolved and particulate matter is chosen for practical reasons at 0.45 μm, but this does not always separate very fine colloidal particles from material truly in solution. Particulate matter (the seston) can remain in suspension for long periods because of turbulence.

2   Phosphate and nitrate are minor constituents and essential nutrients. They are extracted from surface water by photosynthesizing phytoplankton to make organic tissue. They may become totally depleted in surface waters where biological production is high, and are known as biolimiting constituents – they limit production because when they are exhausted, production ceases. When the organisms are consumed or when they die and decompose, the nutrients are returned to the water column (re-mineralized). The molar ratio of N to P in both seawater and organic tissue is about 15 : 1. Silica is also a biolimiting nutrient, but is used only to make the hard parts of some planktonic organisms. The skeletal remains dissolve only slowly as they sink into deep water after death, and can accumulate in sediments on the sea-floor.

3   Carbon is essential to all life, but is so abundant in seawater that its involvement in biological production makes only a small difference to its concentration. Calcium is used to make calcium carbonate skeletons and shells, but like carbon it is so abundant that its concentration is little affected. Carbon and calcium are bio-intermediate constituents. Bio-

unlimited constituents are those whose concentrations are unaffected by biological activity.

4 The four principal atmospheric gases are nitrogen, oxygen, argon and carbon dioxide. Carbon dioxide is the most soluble gas in seawater, but occurs in solution mostly *not as a gas* but as bicarbonate and carbonate ions, the main forms of total dissolved inorganic carbon (ΣCO_2) in seawater. Concentrations of gases in surface waters are determined by their individual solubilities at the prevailing temperature and their atmospheric partial pressure. The solubility of gases decreases with increased temperature and salinity, and increases with pressure. Diffusion rates across the air–sea interface are increased in stormy weather, and dissolved gases are carried to deeper levels mainly by turbulent diffusion.

5 Oxygen is supersaturated in surface waters. The compensation depth at the base of the photic zone can be defined as the depth at which the amount of oxygen used (or carbon 'burnt' or dissipated) in respiration is equal to the amount of oxygen liberated (or carbon fixed) by photosynthesis. Below the photic zone, respiration uses up available oxygen and an oxygen minimum layer develops at a depth of a few hundred metres. Deep water is richer in oxygen because of cold well-oxygenated water sinking in polar regions.

6 The concentration of total dissolved inorganic carbon (ΣCO_2) increases with depth because CO_2 is used during photosynthesis (and formation of calcium carbonate) and released again during respiration (and dissolution of calcium carbonate); and the solubility of CO_2 (and of $CaCO_3$) is increased by increased pressure. ΣCO_2 concentrations are an important factor in controlling the pH of seawater, which is mostly within the range of 7.7 ± 0.2, being greater at the surface (less acid) than at depth (more acid). Many minor gases in seawater are produced by biological activity and are supersaturated in surface layers, so they have a net flux from sea to air.

7 Rainwater is a dilute version of seawater, because aerosols carry marine salts into the atmosphere where they provide nuclei for rain formation. The dissolved constituents in river water result from rock weathering and are dominated by calcium and bicarbonate ions, whereas seawater and rainwater are dominated by sodium and chloride ions. The chloride and much of the sodium in river water come from recycled sea salt. Sodium balance calculations show that most of the dissolved constituents of seawater can be accounted for by rock weathering, but some cannot, notably chloride, bromide and sulphate. These are excess volatiles whose main source is probably volcanic gases. Hydrothermal activity in the ocean basins is an important additional source of some constituents of seawater.

8 The ocean is generally believed to be in a chemical steady state: rates of input and removal of dissolved constituents are in long-term balance. Residence times of dissolved constituents range from about 100 million years down to 1 000 years or less, and there is a very rough correlation between concentration and residence time. The residence time of water in the oceans is about 4 000 years and the average stirring (or mixing or turnover) time is of the order of 500 years.

9 Dissolved constituents in seawater are mostly in ionic form, and all ions are surrounded by a hydration sphere. Many ions form ion pairs or ionic complexes, in which the hydration spheres are more or less merged. Several of the major constituent ions form ion pairs; two of the most abundant pairs are $MgSO_4$ and $MgCO_3$.

10   Minor and trace elements in seawater generally have short residence times. Speciation is important, because several elements have more than one valency and can occur in more than one form. Different forms have different solubilities, and some may be removed by adsorption and scavenging, precipitated by redox reactions, or co-precipitated with insoluble complexes of other elements. Many are involved in biological processes and become greatly enriched in the tissues of marine organisms, so that organic-rich sediments may also be rich in trace elements. Some are biolimiting micro-nutrients; others show bio-intermediate behaviour.

11   *Carbonate, alkalinity and pH:* Some marine organisms use calcium carbonate to form their hard parts, which redissolve when the organisms die and sink into deep water. The depth at which dissolution begins is called the lysocline; the depth at which little or no calcium carbonate remains is called the carbonate compensation depth or CCD. Most dissolution occurs at the sea-bed, so the CCD is a sort of 'snowline'. $CaCO_3$ may be in the form of the less common and less stable aragonite, or the more abundant and more stable calcite. The lysocline and CCD are deeper for calcite than for aragonite. The concentration of calcium is greater in deep than in surface waters because of the dissolution of calcium carbonate. So is that of total dissolved carbon (ΣCO_2), partly because of the dissolution of calcium carbonate, but chiefly because of the decomposition of organic tissue through consumption and respiration. To a good first approximation:

$$[\Sigma CO_2] = [HCO_3^-] + [CO_3^{2-}] \qquad (6.10)$$

Alkalinity (A) is the combined negative charge due to bicarbonate and carbonate ions in seawater, expressed as molar concentrations. It is determined by titration with acid. It can also be defined as the excess of total major cations over total major anions (other than bicarbonate and carbonate), in molar 'charge-equivalent' terms.

Remember:

- A is changed *only* by formation and dissolution of calcium carbonate ($CaCO_3$).
- A is *not* changed by formation and decomposition of organic matter.
- Formation and decomposition of organic matter changes *only* $[\Sigma CO_2]$.
- Formation and dissolution of $CaCO_3$ change $[\Sigma CO_2]$ as well as A.

(Figure 6.15 summarizes these four points.)

$[\Sigma CO_2]$ increases more with depth than does A, so from

$$A - [\Sigma CO_2] = [CO_3^{2-}] \qquad (6.11)$$

the concentration of carbonate ions decreases with depth; and from:

$$[H^+] = K \frac{[HCO_3^-]}{[CO_3^{2-}]} \qquad (6.15)$$

deep water is generally more acid (lower pH) than surface water (higher pH); and skeletal remains formed of calcium carbonate dissolve as they sink into deep water.

Now try the following questions to consolidate your understanding of this Chapter.

QUESTION 6.18   Classify each of the dissolved constituents in Figure 6.19 in the (a) biolimiting; (b) bio-intermediate; or (c) bio-unlimited category.

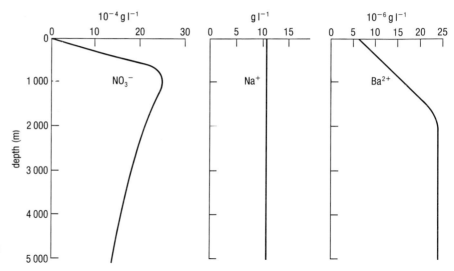

Figure 6.19 Concentration profiles for three seawater constituents (for use with Question 6.18). Note that concentrations are normalized to a salinity of 35.

QUESTION 6.19   The rate of removal of dissolved manganese (Mn^{2+}) from seawater to deep-sea sediments is an order of magnitude greater than the rate of supply of dissolved Mn^{2+} to the ocean by rivers.

(a)   Which of these two rates gives the shorter residence time?

(b)   What additional source of dissolved Mn^{2+} could supply the shortfall between river source and sediment sink in the oceanic manganese budget?

QUESTION 6.20   Under anoxic conditions, would you expect manganese to be in the form of soluble Mn^{2+} ions or insoluble oxide MnO_2?

QUESTION 6.21   Which of the following statements (a)–(h) are true, and which are false?

(a)   Most of the nitrogen dissolved in seawater is in the form of nitrate ions.

(b)   Dissolved gases are conservative constituents of seawater.

(c)   The oxygen minimum layer is below the (oxygen) compensation depth.

(d)   The principal factor which determines the (oxygen) compensation depth is the concentration of dissolved oxygen.

(e)   The greater the average concentration of a dissolved constituent in river water relative to that in seawater, the shorter the residence time of that constituent (assuming rivers to be the main source).

(f)   If Na^+ and Ca^{2+} ions have similar ionic size, then the Na^+ ion will have a larger hydration sphere relative to its size.

(g)   The fact that evaporation is used in the commercial extraction of salts from seawater provides evidence that most major constituents are undersaturated in the seawater solution.

(h)   A large influx of highly acid effluent (or a heavy downpour of acid rain) would tend to expel CO_2 from seawater.

QUESTION 6.22   Figure 6.15 and Table 6.6 show that $[\Sigma CO_2]$ is greater in the deep Pacific than in the deep Atlantic. The same is true of the alkalinity (Figure 6.15 and Table 6.6). What does that imply about the $Ca^{2+}:S$ ratio in deep waters of the two oceans?

CHAPTER 7 | SEAWATER AND THE GLOBAL CYCLE

The role of the ocean in the global cycling of elements is illustrated in Figure 7.1 and may be summarized as follows:

1   Weathering of rocks supplies dissolved constituents to the oceans, aided both by the dissolution of minerals in rocks of the oceanic crust during hydrothermal circulation, and by the supply of chloride, sulphate and other excess volatiles from volcanic gases.

2   Most dissolved constituents have residence times much longer than oceanic stirring times (Figure 6.11) and may be cycled repeatedly within the main body of the oceans, especially by participating in biological reactions. There can also be exchanges across the sea-bed and the air–sea interface.

3   Dissolved constituents are eventually removed from the seawater solution into sediments and rocks by the processes of reverse weathering. These include adsorption and scavenging of minor and trace elements, formation of skeletal material, diagenetic reactions with sediments, preservation in anoxic environments, and reactions during hydrothermal activity.

4   Sediments and rocks are removed from the oceanic environment either by direct uplift above sea-level or by **subduction** into the Earth's mantle, at **destructive plate margins**. Uplift brings sediments and rocks directly back into the weathering environment; subduction eventually returns them to the crust via magmatic processes which form igneous rocks and release volcanic gases (including excess volatiles).

Figure 7.1   Over periods of millions of years, the oceans behave as a well-mixed tank. The input and output processes are either external (Sun-driven: river flow, photosynthesis); or internal (Earth-driven: reactions at mid-oceanic ridges, uplift and subduction of oceanic crust).

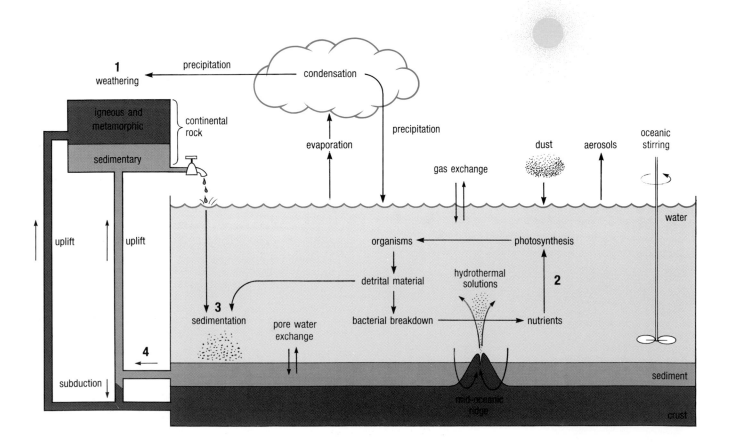

The excess volatiles are thus part of the global cycle. A small proportion is probably truly juvenile or primordial (i.e. derived from deep within the Earth, where it has resided from the beginning); but most of these volatiles have been circulating through the system summarized in Figure 7.1 for perhaps thousands of millions of years.

QUESTION 7.1 Carbon dioxide and methane are also volcanic gases. While a proportion of both gases may be juvenile in origin, most is probably recycled. Can you suggest how this could happen?

Geological evidence supports the proposition made in Section 6.2.5 that the composition of seawater has not changed significantly for at least several hundred million years. Marine sedimentary and igneous rocks of all geological ages are compositionally very similar to their modern equivalents; this applies not only to the major components of limestones, sandstones and shales, but also to concentrations of minor and trace elements. For example, evaporites older than 2 500 million years retain evidence that sodium chloride was the principal salt to be precipitated. Proportions of elements such as copper, zinc and uranium in organic-rich (black) marine shales from ancient geological periods are similar to those in comparable sediments being deposited now in the Black Sea. The chemical steady state thus appears to be a fundamental characteristic of the oceans. Rates of input and removal of most dissolved constituents must be generally in balance, which is why we are able to calculate residence times. Long-term stability of composition does not mean eternal constancy, however, and there must have been some changes in seawater composition with time.

7.1 A SHORT HISTORY OF SEAWATER

The most significant influence on the surface environment of the early Earth must have been the composition of the atmosphere, which was dominated by nitrogen and carbon dioxide. There was no free oxygen in the primitive atmosphere, which probably also contained methane (another de-gassing product of the early Earth; cf. Chapter 1), and there may have been local reduction of nitrogen to ammonia. Laboratory experiments in the 1950s showed that amino acids (the building blocks of protein) could have been naturally synthesized from these gases in solution in seawater. The necessary energy could have been supplied by lightning discharges and by ultraviolet radiation, which could penetrate the atmosphere to the Earth's surface in the absence of atmospheric oxygen and hence of an ozone layer. With the discovery of hydrothermal vents in the oceans during the 1970s, it was suggested that organic molecules necessary for the development of early life forms could also have originated in the deep sea. Hydrothermal vents would provide an ideal environment: plenty of hot water and an abundance of raw materials. The Earth was hotter early in its evolution than it is now, and the hydrothermal environment was more widespread. Some meteorites contain organic molecules, including hydrocarbons and amino acids, so the 'organic soup' of the primordial Earth may even have been derived from space. Whatever the origin of life on Earth, however, the earliest life forms preserved in the fossil record are about 3.5 billion years old, and they were primitive blue–green algae (Cyanobacteria), requiring sunlight for photosynthesis.

The composition of seawater on the early Earth must also have been different from what it is now. Dissolved constituents such as bicarbonate and sulphate among the anions, and iron and manganese among the cations, must have been present in very different proportions. The concentration of HCO_3^- was probably second only to that of Cl^-, because sulphur would mostly have been in the form of relatively insoluble sulphide rather than soluble sulphate. The reduced forms of iron and manganese (Fe(II) and Mn(II)) are both readily soluble, and as these elements are major components of crustal rocks (Table 3.2) they would have been more abundant in seawater than they are now.

The ratio of atmospheric CO_2 to O_2 gradually decreased with time, as photosynthesizing organisms fixed carbon in organic tissue and released oxygen (reaction 6.3). A contributory factor was the accumulation of calcium carbonate, mainly *stromatolites* precipitated as mats by shallow water algae which were in existence at least three billion years ago (animals with calcareous skeletons did not evolve until about 600 million years ago).

With an oxidizing atmosphere and a progressive decrease in CO_2, both atmosphere and ocean approached their present compositions ever more closely. The temperature-buffering effect of the oceans (Section 2.1) has kept the Earth's surface tolerable for life, but average surface temperatures have fluctuated appreciably. There is good geological evidence that the Earth's surface environment is normally characterized by ice-free poles and relatively gentle poleward temperature gradients. In short, the present-day Earth is in an atypical condition: we may still be in the Pleistocene Ice Age, albeit enjoying the relative warmth of an interglacial interval. The previous major ice age occurred about 300 million years ago, during the Permo-Carboniferous Period.

QUESTION 7.2 If higher average temperatures than now characterized the Earth's surface, and the poles were ice-free, what implications might such conditions have for the deep circulation?

It is worth amplifying the answer to Question 7.2: in an ice-free world, lower dissolved oxygen concentrations in deep water would also mean less decomposition and more preservation of organic tissue in sediments. There would be less rapid turnover of nutrients than now, because more organic tissue, containing the nutrients, was being preserved in sediments.

7.1.1 THE SPECIAL CASE OF CO$_2$

Carbon is the element which forms the basis of all life (Section 6.1.2) and carbon dioxide is the form in which it is used for photosynthetic primary production. Early in the Earth's geological history, CO_2 was about a thousand times more abundant than in the pre-industrial atmosphere of *c.* 200 years ago (before the increase in fossil fuel burning). Its partial pressure would have been nearly 0.3 atmosphere, which is more than the present-day partial pressure of oxygen. In volumetric terms, CO_2 may have been second only to nitrogen in the atmosphere of the early Earth. Greater partial pressure of CO_2 in the atmosphere would probably mean that $[\Sigma CO_2]$ in seawater was greater, too (reaction 6.2).

QUESTION 7.3 Does that mean the early ocean was more acid (lower pH) than it is now?

Some authorities estimate that early in the Earth's history the solar luminosity was about 25% less than it is now, so that the Earth received only about three-quarters of the present-day solar radiation. If that is correct, then solar luminosity must have increased with time, as atmospheric CO_2 has progressively decreased. As outlined in Section 2.1, carbon dioxide is the main contributor to the atmospheric greenhouse effect, along with water vapour. The more CO_2 in the atmosphere, the greater the effect.

If CO_2 levels had not fallen as solar luminosity increased, what would the Earth's surface be like now?

Very hot: rather like Venus is today. Unlike Venus, however, the early Earth was just far enough from the Sun for liquid water to exist at the surface, in rivers, lakes and seas. As soon as carbon dioxide dissolved in the water, its abstraction from the atmosphere could begin, through rock weathering (e.g. reaction 6.5) and through accumulation of organic matter and calcium carbonate in sediments. Much carbon is now locked up in crustal rocks, as well as in the biosphere and in fossil fuels (Table 7.1). Carbon continues to circulate through the global cycle, but the amount stored in the various reservoirs changes rather little. (Although the fossil fuel 'bank' is being rapidly depleted by human activities, this is a relatively small reservoir of carbon.)

Table 7.1 Amounts of carbon in various reservoirs (10^{12} tonnes CO_2 equivalent).

Reservoir	Approximate quantity
Atmosphere	3
Biomass (living matter)	3
Disseminated organic carbon in soils & sediments	125 000
Oceans and freshwater (in solution)	140
Carbonate sediments	150 000
Fossil fuels	35

The decline in atmospheric CO_2 was progressive, but it may not have been regular. Figure 7.2 provides evidence that considerable short-term fluctuations of atmospheric CO_2 concentration have occurred during the past couple of hundred thousand years at least.

Perhaps the most interesting feature of Figure 7.2 is not the fluctuations of atmospheric CO_2 themselves, but the strong correlation with surface temperature.

Is it a positive or a negative correlation?

The correlation is positive, that is, as temperature increases, so does the atmospheric CO_2 concentration. The deglaciation (warming) events commencing about 140 000 years and 15 000 years ago are particularly obvious. The fluctuations in Figure 7.2 are consistent with what might be expected from simple considerations of solubility and temperature alone. As CO_2 is more soluble in cold than in warm water, its atmospheric concentration should be less during glacial periods of lower mean temperature than interglacial periods of higher mean temperature.

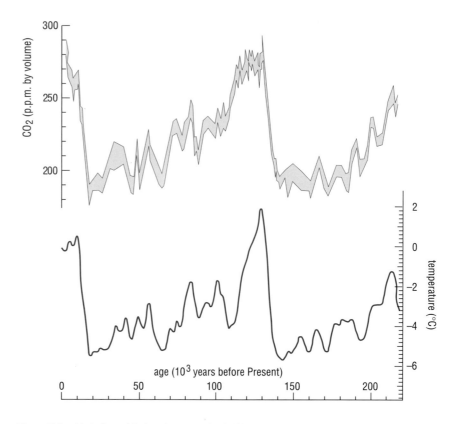

Figure 7.2   Variation with time in atmospheric CO_2 concentrations, determined from air bubbles trapped in an ice core from Vostok in eastern Antarctica (grey curve – width of shaded area corresponds to measurement error); along with the atmospheric temperature at the surface, inferred from measurements of the deuterium/hydrogen isotopic ratio in H_2O (red curve).

The full picture is more complex than that, however, and there has been considerable debate about whether changes in the atmospheric CO_2 concentration are a response to the temperature fluctuations or a cause of them. Several hypotheses have been proposed, relating changes in atmospheric CO_2 concentration to changes in biological productivity, in rates of terrestrial rock weathering, in sea-level, and in the circulation of surface and deep current systems (including the relative importance of deep water mass formation in high latitudes, and of upwelling regions where biological production is high). As these factors are all interrelated, the resulting models are complex and no clear cause-and-effect relationships have emerged so far. It is reasonably certain, however, that varying CO_2 concentrations in the atmosphere are not the main cause of temperature fluctuations – they may reinforce climatic trends, once established, but are not likely to initiate them.

This is a suitable point at which to digress briefly into some external factors which influence the Earth's climate.

7.1.2   CLIMATE AND THE EARTH'S ORBIT

Seasonal variations in insolation were briefly described in Section 2.2. Figure 7.3 provides a more detailed picture of the way insolation varies with latitude and season.

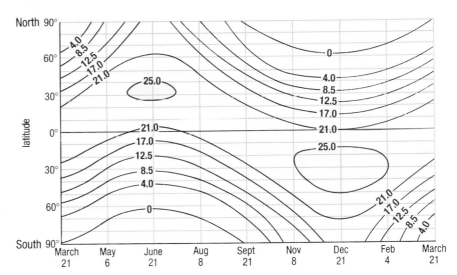

Figure 7.3 Seasonal variation of daily insolation (in 10^6 J m^{-2} at the Earth's surface), assuming 30% reflection from the top of the atmosphere (cf. Section 2.1). Values are highest in mid-latitudes because of long daylengths in summer.

QUESTION 7.4 Which parts of the Earth's surface receive the greatest insolation (a) at any one time, (b) on average over the year as a whole?

In answering Question 7.4, you may also have noticed that Figure 7.3 shows mid-latitudes in the Southern Hemisphere to receive more solar radiation than corresponding latitudes in the Northern Hemisphere, during their respective summer seasons. This is because the Earth is in an elliptical orbit around the Sun, so its distance from the Sun changes through the year. The positions on the orbit closest to and furthest from the Sun are called the **perihelion** and **aphelion** respectively.

So, in what part of the year is the Earth at perihelion, according to Figure 7.3?

Insolation is greatest during the southern summer, so the Earth must be at perihelion then – early January in fact, close to the December solstice (cf. Figure 2.2).

The geometry of the Earth's orbit round the Sun changes with time, and this affects the seasonal and latitudinal distribution of the insolation received by the Earth each year. We can identify three main cyclical changes in the orbital configuration, having periods of about 110 000 years, 40 000 years, and 22 000 years. The longest cycle (Figure 7.4(a)) affects the *eccentricity* (or *ellipticity*) of the Earth's orbit, which changes in shape from elliptical to nearly circular and back again because of the varying gravitational attraction of the Sun, the Moon and the planets (notably Jupiter and Saturn).

In what major respect would Figure 7.3 look different if the Earth's orbit were nearly circular?

Insolation in both hemispheres would be about the same, because perihelion and aphelion become irrelevant when the orbit is circular. The degree of eccentricity of the Earth's orbit at present can be judged from the fact that the Sun–Earth distance is 1.521×10^8 km at aphelion, 1.471×10^8 km at perihelion.

The two shorter cycles (Figure 7.4(b)) involve the orientation of the Earth's axis, which varies so that the direction in which the North Pole points traces a circle in the sky (i.e. what we call the Pole Star has not always been an accurate indication of North). The time taken for a full circle to be traced in the sky – a phenomenon manifested by the *precession of the equinoxes* – is about 22 000 years. The precession of the equinoxes causes the seasonal position of the Earth in its (elliptical) orbit to change: the December solstice has not always occurred near perihelion, though it did 22 000 years ago and will again 22 000 years in the future. At the same time, the *tilt* of the Earth's axis (which is presently at about 23.4°) changes between 21.8° and 24.4° and back again, with a periodicity of ~40 000 years. The greater the tilt of the axis, the greater the difference between summer and winter: at present the tilt is decreasing, so summers are very gradually become cooler and winters are very gradually becoming warmer.

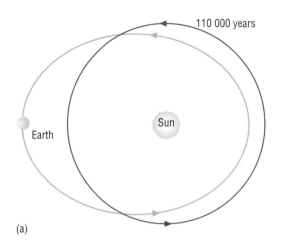

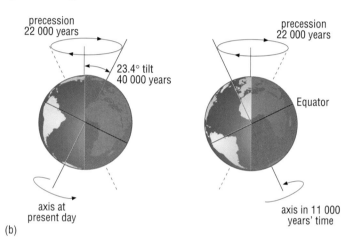

(a) (b)

Figure 7.4   The component Milankovitch cycles. (a) The eccentricity cycle (plan view not to scale). When the orbit is nearly circular, the Sun is at the centre of the circle; when the orbit is elliptical, the Sun is at one of the foci of the ellipse. (b) and (c) show tilt and precession cycles. (b) shows the Earth in the present stage of its tilt and precession cycles, at the December solstice (northern winter), when it is near perihelion (see text), and the tilt is 23.4°. (c) When the Earth is near perihelion 11 000 years hence, its axis will tilt in the opposite direction (because it will be half-way round the 22 000-year precession cycle), and the angle of tilt will be approaching the minimum value of 21.8°. Maximum insolation will be experienced during Northen Hemisphere summer, the opposite of the present situation (Figure 7.3, Question 7.4 and related text).

These three periodicities, *c.* 110 000 years, *c.* 40 000 years and *c.* 22 000 years, are called **Milankovitch cycles**, after the scientist who became famous for recognizing (in the 1920s and 1930s) that because they affect solar insolation patterns they could also be an agent of global climatic change[*]. And so it has proved. Fluctuations with these periodicities can be discerned in temperature records determined from the oxygen isotope ratios measured in the calcareous skeletons of plankton preserved in deep-sea sediments.

Over the past million years or so of the Earth's history, the interval between major glaciation/deglaciation events corresponds most closely with the 110 000-year eccentricity cycle, while smaller fluctuations can be correlated with the shorter cycles (cf. Figure 7.2).

However, the effects of the *c.* 110 000-year orbital eccentricity cycle are generally considered insufficient by themselves to bring about the major changes of surface temperature and ice cover implied in compilations such as Figure 7.2. Interactions between the oceans, the ice-sheets and the atmosphere, and the characteristics of the ice-sheets themselves, must also be considered. A further complication is that the *c.* 110 000-year eccentricity cycle, the effects of which seem so obvious on Figure 7.2, appears to be a characteristic of only about the last million years of the Earth's history. It is

[*] In fact, the first to recognize these relationships was the 19th century scientist, James Croll, and these cycles are therefore sometimes referred to as Milankovitch–Croll cycles.

less obvious in records of earlier climatic fluctuations preserved in sediments, which are dominated by the *c.* 40 000-year tilt cycle back to about two-and-a-half million years ago, the start of the Pleistocene Ice Age.

A fascinating suggestion to account for this difference is that the rapid tectonic rise of mountains in the Himalayas and western North America affected the circulation in the upper atmosphere in such a way as to strengthen the influence of the 110 000-year eccentricity cycle – and it provides an example of how the solution of modern global problems requires information from many different fields of science.

7.2   A LOOK AHEAD

Figure 7.5 shows how the content of CO_2 in the atmosphere has increased since the Industrial Revolution. In recent years this increase has accelerated, due partly to greater industrial activity and partly to greatly increased deforestation and land clearance for urban, industrial and agricultural development. Many people are now aware of the likely enhanced greenhouse effect of this accelerated increase (Section 2.1): atmospheric and surface warming on a time-scale of decades, with consequences such as melting ice-caps and rising sea-levels. Mean global surface temperature has increased by about 0.5 °C since the late 19th century, and mean sea-level has risen some 10–15 cm in the same period, partly because of melting ice, but also because of thermal expansion of the top few hundred metres of the seawater column. By about the year 2030, mean temperature and sea-level may have risen further by similar amounts or even more. These changes

Figure 7.5   (a) Increase in atmospheric CO_2 since medieval times determined from air trapped in Antarctic ice. (Rectangle at upper right is the post-1950 record, see (b).)

(b) Increase in atmospheric CO_2 measured at the Mauna Loa observatory, Hawaii, along with the increase predicted on the basis of fossil fuel combustion.

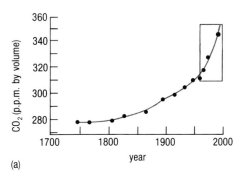

(a)

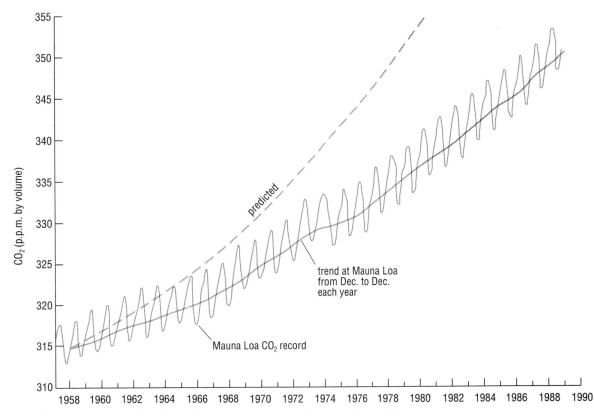

(b)

cannot be unequivocally attributed to an enhanced greenhouse effect resulting from human activities, but there is plenty of evidence for a link between atmospheric temperature and CO_2 concentration (e.g. Figure 7.2).

QUESTION 7.5 How do present-day atmospheric concentrations of CO_2 (Figure 7.5) compare with those of 130 000 years ago (Figure 7.2)?

The remarkable thing about this comparison is that the increase of *c.* 70 p.p.m. from the year 1800 to the present has occurred in less than 200 years. At the start of the last deglaciation, a comparable rise took about 5000 years. The rate of increase of atmospheric CO_2 is probably greater now than it has been at any time in the Earth's history. The difference between the 'observed' and 'predicted' curves in Figure 7.5 shows that not all of the CO_2 released by human activities has stayed in the atmosphere. Some of the excess is probably used up by increased rates of terrestrial photosynthetic production and rock weathering; and about a third has dissolved in the oceans, where it is available for greater primary production.

The biosphere thus appears to be counteracting the artificial increase in atmospheric CO_2 by acting as a sink for some of it, so that the temperature rise is less. Earlier, you read that the principal cause of the progressive fall in the atmospheric $CO_2 : O_2$ ratio was liberation of oxygen and removal of CO_2 to sediments. Relationships of this kind have led to the suggestion that the surface of our planet is maintained as a life-supporting environment by biological activity, via a variety of feedback mechanisms. This is the cornerstone of the **Gaia Hypothesis**, which was put forward in the early 1970s, but only came to prominence in the mid-1980s.

It is tempting to see a biological feedback mechanism in correlations such as that in Figure 7.2: increased photosynthetic primary production during warm climatic periods removes large amounts of atmospheric CO_2, its concentration decreases and the climate cools; primary production declines in the cooler conditions, CO_2 accumulates in the atmosphere, and the climate warms again. However, the variations in Figure 7.2 are too precisely synchronous for this to be the sole (or even the principal) explanation, and other factors must be involved. For example, sulphate aerosols introduced into the atmosphere by large volcanic eruptions (Section 6.1.3) can increase the Earth's albedo, both directly through scattering of solar radiation and indirectly through cloud formation (Section 2.1). Thus, mean global temperature was slightly but measurably lower for a couple of years following the eruption of Mount Pinatubo, which emitted huge tonnages of sulphur dioxide. It has also been suggested that increased amounts of sulphate aerosols resulting from industrial activities of the past few decades could have the same effect, counteracting some of the global warming caused by rising concentrations of greenhouse gases.

Nonetheless, a great deal of research effort continues to be put into establishing the extent to which marine phytoplankton take up excess CO_2 and so might at least slow down the increase in atmospheric concentration and hence the rate of global warming. International and national programmes initiated to investigate these and related questions have included the Joint Global Ocean Flux Study (JGOFS) and the somewhat broader World Ocean Circulation Experiment (WOCE).

Scientists have even considered ways of artificially increasing marine biological production to remove more CO_2 from the atmosphere. For example, dissolved iron has been identified as a possible biolimiting micronutrient (cf. Figure 6.17), because in some productive regions of the ocean, nitrate and phosphate are still available in surface waters where concentrations of iron are below detection limits. In the early 1990s, this led to proposals that the upper ocean might be 'fertilized' with controlled quantities of dissolved iron, to stimulate phytoplankton production. A large area of iron-deficient surface ocean in the eastern equatorial Pacific was experimentally 'dosed' with iron solution. Phytoplankton production did increase somewhat but the increase was shorter-lived and smaller than anticipated. While the results indicated that iron limitation can control rates of phytoplankton production, many scientists have argued that attempts at 'planetary engineering' by measures such as iron fertilization are misguided. It would make more sense to try to reduce the CO_2 emissions themselves.

Climatic instabilities

The past 10 000 years of the Earth's history appear to have been characterized by particularly stable climatic conditions. Fluctuations such as the medieval warm period and the Little Ice Age of the 17th and 18th centuries represented departures of less than 1 °C from present-day global mean surface temperatures.

However, throughout most of the preceding 200 000 years or so, the climate appears to have been much more changeable. Figure 7.6 summarizes the evidence. It compares data from the Antarctic Vostok ice core (Figure 7.2) with a record obtained by drilling in the central 'Summit' region of the 3-km-thick Greenland ice-cap during the GReenland Ice-core Project, GRIP), initiated in the early 1990s.

What is the most notable difference between the two records in Figure 7.6 (other than the sizes of the ratios themselves, which result from the location of the ice-caps at opposite poles)?

The GRIP record is much 'spikier', suggesting that fluctuations of surface temperature were generally more frequent and extreme in the Northern than in the Southern Hemisphere, up to about 10 000 years ago. Temperatures fluctuated by several degrees on time-scales as short as centuries and perhaps even decades. It is not altogether surprising that these changes should be so marked in the Northern Hemisphere (often called the 'Land Hemisphere') because seasonal (and interannual) variations of temperature are much greater on land than over the oceans (Figures 2.1 and 2.3).

Data such as those in Figures 7.2 and 7.6, together with information from deep-sea sediment cores, suggest that the relatively stable climate of the Holocene interglacial has been the exception rather than the rule, at least during the last couple of hundred thousand years of Earth history. Even quite small perturbations of the ocean–atmosphere system could trigger a 'climatic flicker', a rapid cooling or warming event. Concerns about global warming no longer focus exclusively on a progressive world-wide (albeit uneven) rise of temperature and on how the effects of this could be ameliorated or countered. The question is now whether rising concentrations of atmospheric CO_2 from burning fossil fuels and deforestation could warm the atmosphere sufficiently to 'flip' the climate into a phase of accelerated warming or cooling, like one of the 'spikes' in Figure 7.6.

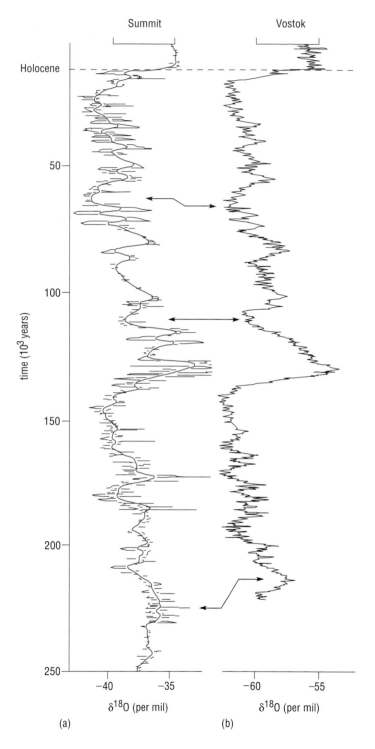

Figure 7.6 Oxygen isotope records from (a) the GRIP 'Summit' ice core, and (b) the Antarctic Vostok ice core (cf. Figure 7.2). The blue curve in (a) results from statistical filtering of the data using a 5 000-year interval. Variations in the oxygen isotope ratios ($\delta^{18}O$) of the ice, expressed in parts per thousand (per mil), provide a measure of atmospheric temperature fluctuations in polar regions during the period represented by the cores, because $\delta^{18}O$ in polar glacier ice is mainly determined by its temperature of formation; the main point is that the larger the (negative) number, the lower the temperature. Note that the relative climatic stability of the past 10 000 years (the Holocene) encompasses most of recorded history, but represents only a small fraction of the records.

Here is just one example of how that might happen: the Gulf Stream and the North Atlantic Drift (Figure 2.11) transport warm and relatively saline water into the Greenland and Norwegian Seas, where the water is cooled and becomes denser and sinks, to form North Atlantic Deep Water, a major component of the thermohaline circulation (Sections 4.1 and 4.3.3). As this water cools and sinks, vast amounts of heat are liberated into the atmosphere, and it is this heat which is largely responsible for the relatively equable climate of north-western Europe.

Global warming could lead to accelerated melting of glaciers and ice-caps in north polar regions, and the resulting melt waters could spread out across the surface of the North Atlantic.

Why would they not simply mix into the main body of water in the North Atlantic, and what effect would a 'lid' of melt water have on production of NADW?

The melt waters would spread out across the surface because of their low density, a consequence of their low salinity. The salinity would be too low for the 'lid' to become dense enough to sink through the much higher salinity seawater beneath. Production of North Atlantic Deep Water would effectively cease, and north-western Europe would no longer benefit from the heat liberated by this process.

Somewhat paradoxically then, the effect of continued global warming could be a period of rapid *cooling*, perhaps exacerbated by increased ice and snow cover which would reflect more solar radiation back to space because of its high albedo (Table 1.2).

At all events, evidence from GRIP cores (and also from cores drilled by the companion Greenland Ice Sheet Project (GISP)) suggests that the climatic system in the North Atlantic region can change rapidly, perhaps even within decades; this could hardly fail to have profound effects on other regions of the globe. It is not yet clear *why* such a change should occur (or even whether it will occur at all). In the meantime, the strong positive correlation between atmospheric concentrations of CO_2 and surface temperature (Figure 7.2) suggests that climatic warming will continue to accompany the exponential increase of carbon dioxide and other greenhouse gases in the atmosphere (Figure 7.5), until something happens to reverse the trend.

7.3 SUMMARY OF CHAPTER 7

1 The oceans are maintained in a chemical steady state by the global cycling of elements through (i) weathering and solution; (ii) precipitation, re-solution and sedimentation; (iii) subduction, uplift, volcanic activity; and (iv) back to weathering. Sediments and rocks of the marine environment have changed little in composition through geological time, which suggests that seawater composition has not changed greatly and that elements have been cycled through the oceans in similar proportions and amounts since early in the Earth's history.

2 Seawater composition may initially have been governed partly by high $CO_2 : O_2$ ratios in the early atmosphere, with greater bicarbonate and lower sulphate concentrations, and greater concentrations of reduced cationic

species (e.g. Fe^{2+} and Mn^{2+}). $CO_2 : O_2$ ratios decreased as carbon was fixed and oxygen released by photosynthesis.

3 The decrease in CO_2 with time was accompanied by increased solar luminosity. The greenhouse effect of atmospheric CO_2 decreased with time, thus helping to maintain an equable surface temperature. The decline in atmospheric CO_2 is progressive, but probably irregular. The typical surface environment of the Earth is one of ice-free poles and higher average temperatures than at present.

4 Fluctuations in the concentration of atmospheric CO_2 are strongly correlated with surface temperature, at least over the past 200 000 years, and probably much longer. Cycles of eccentricity of the Earth's orbit, changes in the angle of tilt of its axis of rotation, and precession of the equinoxes (*c.* 110 000, *c.* 40 000 and *c.* 22 000 years respectively – the Milankovitch cycles), are probably the major causes of climatic variation. Atmospheric CO_2 appears to reinforce trends of changing global temperature, but is probably not a major factor initiating them.

5 As CO_2 levels now increase again because of human activities, the enhanced greenhouse effect is predicted to lead to melting ice-caps and rising sea-levels. Some excess CO_2 is taken up by terrestrial and marine photosynthetic activity. The notion that the biosphere helps to maintain the Earth's surface in a life-supporting condition is the basis of the Gaia Hypothesis.

6 Ice core records suggest that the climate of the past 10 000 years has been untypically stable. Much of the preceding 200 000 years was characterized by phases of abrupt and rapid climatic warming or cooling ('flickers'), especially in the Northern ('Land') Hemisphere. Temperatures seem to have risen or fallen several degrees on time-scales as short as decades. Global warming caused by increased concentrations of CO_2 (and other greenhouse gases) in the atmosphere could 'flip' the climate system into one of these phases of rapid change.

Now try the following questions to consolidate your understanding of this Chapter.

QUESTION 7.6 (a) It is sometimes said by marine chemists that weathering consumes acids and releases alkalinity into solution, whereas reverse weathering consumes alkalinity and releases acids into solution. Recalling the definition of alkalinity (Section 6.3.2), can you interpret this somewhat cryptic description, with the help of equation 7.1?

$$\underset{\substack{\text{(hydrogen} \\ \text{ions in} \\ \text{solution: acid)}}}{H^+(aq)} + \underset{\substack{\text{(calcium} \\ \text{carbonate)}}}{CaCO_3(s)} \underset{\substack{\text{reverse} \\ \text{weathering}}}{\overset{\substack{\text{weathering}}}{\rightleftharpoons}} \underset{\substack{\text{(calcium and bicarbonate} \\ \text{ions in solution:} \\ \text{alkalinity)}}}{Ca^{2+}(aq) + HCO_3^-(aq)} \qquad (7.1)$$

(b) Does the dissolution of $CaCO_3$ in the deep ocean consume or release alkalinity?

(c) Can the solution of excess (anthropogenic) atmospheric CO_2 in the oceans *by itself* affect the alkalinity of seawater?

QUESTION 7.7   If the oceans remain in a steady state and the rate of input of a dissolved constituent increases, what must happen (a) to the rate of removal, (b) to the residence time of that constituent?

QUESTION 7.8   (a)   Bearing in mind the effect of pressure on reaction 6.2 and the fact that gases are more soluble in cold than in warm water (Section 6.1.3), explain why areas of upwelling in low latitudes tend to be associated with a net flux of CO_2 from sea to air, whereas areas of formation of deep water masses (Section 4.1) tend to be net sinks for atmospheric CO_2.

(b)   What would counteract the tendency for areas of upwelling to be net sources of CO_2 to the atmosphere?

QUESTION 7.9   (a)   Why might cessation of NADW formation lead to a reduction in the rate of removal of CO_2 from the atmosphere?

(b)   Could a 'lid' of low salinity water over the North Atlantic also inhibit biological removal of CO_2 by the plankton?

APPENDIX   CONVERSIONS BETWEEN pH AND [H+]

This Appendix is intended as a reminder about how to manipulate the relationship between pH and the hydrogen ion concentration, [H+].

By definition

$$pH = -\log_{10}[H^+]$$

There is no difficulty in converting whole number values of pH into [H+]. A pH of 8, for example, means that $[H^+] = 10^{-8}$ mol l^{-1}; a pH of 5 means that $[H^+] = 10^{-5}$ mol l^{-1}.

The problem comes when we wish to convert a pH value of, say, 7.4 into [H+] in mol l^{-1}.

The first thing is to write $[H^+] = 10^{-7.4}$ mol l^{-1}; this tells you straight away that [H+] must lie between 10^{-7} and 10^{-8} mol l^{-1}. So you know what to aim for.

The simplest procedure is set out below.

From the definition of pH

$$7.4 = -\log_{10}[H^+]$$
or
$$-7.4 = \log_{10}[H^+]$$

This next step is the most important one:

$$-7.4 = -8 + 0.6$$

The logarithm is now in two parts that can be treated separately. You know from the preamble that the −8 part becomes 10^{-8}, but what about the +0.6? From calculator or tables, the antilogarithm of 0.6 is 4, to the nearest whole number, so

$$[H^+] = 4 \times 10^{-8} \text{ mol l}^{-1}$$

which is in agreement with the result we anticipated from inspection.

Using exactly the same procedure you can now do another example for yourself.

The pH of a seawater sample is 8.3. What is its hydrogen ion concentration, [H+], in mol l^{-1}?

By definition

$$8.3 = -\log_{10}[H^+]$$

$$-8.3 = \log_{10}[H^+]$$

By inspection, you can see that $[H^+] = 10^{-8.3}$ mol l^{-1}, so you know it must lie between 10^{-8} and 10^{-9}:

$$-8.3 = -9 + 0.7$$

The −9 part of your 'composite' logarithm becomes 10^{-9} and the antilogarithm of 0.7 is close to 5, so the hydrogen ion concentration of seawater with pH of 8.3 is

$$[H^+] = 5 \times 10^{-9} \text{ mol l}^{-1}$$

The reverse procedure, the conversion of [H+] into pH, will of course present few problems when simple integers are involved. When [H+] is 10^{-6} mol l^{-1}, the pH is obviously 6.

But what about a hydrogen ion concentration of 7×10^{-8} mol l^{-1}?

The first thing to do is inspect the number and realize that it lies between 10^{-7} and 10^{-8}. So the pH will lie between 7 and 8.

From the definition of pH, you know that you must take the logarithm of 7×10^{-8}, and you know from the nature of logarithms that you need only to *add* the logarithms of these two numbers. The logarithm of 10^{-8} is -8, and the logarithm of 7 is 0.85 to two decimal places. So the logarithm of this value of [H+] is

$$-8 + 0.85 = -7.15$$

By definition, then, the pH in this case is 7.15. We said that from inspection you would expect it to lie between 7 and 8, and you can see from the size of the number that [H+] must lie closer to 10^{-7} than 10^{-8}, and similarly the pH is closer to 7 than to 8.

Now try another one. What is the pH of a sample of seawater having a value for [H+] of 6×10^{-9} mol l^{-1}?

By inspection, that value lies between 10^{-8} and 10^{-9}, and nearer to 10^{-8} than 10^{-9}, so the pH will be between 8 and 9, and probably nearer to 8.

The logarithm of 6×10^{-9} becomes

$$-9 + 0.78 = -8.22$$

By definition, the pH is therefore 8.22.

SUGGESTED FURTHER READING

CHESTER, R. (1990) *Marine Geochemistry*, Unwin Hyman. Comprehensive undergraduate text that reviews sources, sinks and cycles involving dissolved constituents in the oceans.

GARRISON, T. (1993) *Oceanography – An Invitation to Marine Science*, Wadsworth Publishing Co. (California). Very well illustrated introductory text aimed at students with little scientific background, proving useful reinforcement of fundamental principles in an interdisciplinary framework.

GROVES, D. (1989) *The Oceans: A Book of Questions and Answers*, Wiley. An entertaining little volume to test the reader's understanding of all aspects of marine science – but beware: the answers are not always right.

HERRING, P. J., CAMPBELL, A. K., AND WHITFIELD, M. (eds) (1990) *Light and Life in the Sea*, Cambridge University Press. A collection of advanced review and research papers dealing with biological and biochemical aspects of underwater optics.

LIBES, S. M. (1992) *An Introduction to Marine Biochemistry*, Wiley. Advanced undergraduate text on the physical, geological and biological processes that affect the behaviour of elements in seawater and help to control its properties.

SCHLESINGER, W. H. (1991) *Biogeochemistry – An Analysis of Global Change*, Academic Press. Interdisciplinary approach dealing with the biological and geological processes that govern the cycling of chemical elements through the oceans.

THURMANN, H. V. (1991) *Introductory Oceanography* (6th edn), Macmillan. Very well illustrated introductory undergraduate text providing an interdisciplinary overview of the marine sciences as a whole.

ANSWERS AND COMMENTS TO QUESTIONS

CHAPTER 1

Question 1.1   This answer is much fuller than any we expect you to give. It should raise further questions in your mind, to which you will find the answers as you read on.

The oceans are *salty* because water is a very good solvent, and rivers bring vast amounts of dissolved salts to the sea each year.

The oceans are *cold* below the surface layers (which nowhere extend deeper than a few hundred metres) because they are heated from the top, and water is a poor conductor of heat with a high specific heat. The main way in which heat is transferred downwards is not conduction, but turbulent mixing (see Section 2.3).

The oceans are *dark* below depths of a few hundred metres, because light is not transmitted far through water.

The oceans *teem with life*, partly because water is an essential and major constituent of all life forms on Earth, and partly because water is a good solvent and seawater is therefore rich in the nutrients essential to living organisms. Many marine animals are *noisy* and as water transmits sound very well, it is used for communication.

The oceans are *never still* because water is a mobile liquid. Waves, tides and currents keep the water in constant motion.

It would also be legitimate to mention the boiling and freezing points of pure water at 100 °C and 0 °C. Liquid water only exists at the surface of the Earth because surface temperatures lie between 0 °C and 100 °C nearly everywhere. (This may be a circular argument: water has high specific and latent heats, so the buffering effect of the oceans has helped to keep surface temperatures of the Earth within this range.)

Question 1.2   The 'maximum density' must increase along the line, with falling temperature and increasing dissolved salt content.

Question 1.3   (a)   Figure 1.3 shows that 336 'units' of water evaporate from the oceans annually (where 1 'unit' = 10^{15} kg). Precipitation in the oceans is 300 units per year, and run-off adds another 36 (note that this figure includes the seepage of groundwater). 300 + 36 = 336, so the processes are in balance.

(b)   The oceans contain 1 322 000 units and the annual input/output rate is 336 units, so:

$$\text{residence time} = \frac{1\,322\,000}{336} \approx 3\,900 \text{ years}$$

(c)   336 units of water evaporate from the oceans annually, and another 64 units evaporate from land. This total of 400 units is the amount that moves through the atmosphere annually, for the evaporation is balanced by precipitation and run-off. Note, however, that there are on average only about 13 units in the atmosphere at any one time, so the residence time of water in the atmosphere is measured in days.

Question 1.4 (a) The first reason is that liquid water is more dense than ice (Table 1.2). The second reason is that dissolved salts increase the density of water (Figure 1.2).

(b) The dissolved salts have lowered the freezing point (and increased the density) of the water that forms the droplets. The salt droplets will therefore 'melt' their way through the bottom of the ice. The older the ice, the longer this process will have been operating.

Question 1.5 (a) Both the specific heat and the latent heats of fusion and evaporation are extremely high. Large amounts of heat are required even to raise the temperature of water, let alone to evaporate it. The enormous volume of oceanic waters thus provides a huge temperature buffer, which confines the range of temperatures at the Earth's surface to a few tens of degrees centigrade.

(b) Pure water reaches its maximum density at 4 °C, well above the freezing point. The density then decreases again from 4 °C to 0 °C (Table 1.2). But in seawater, this does not happen; density increases right down to the freezing point (Figure 1.2), which is very important for the control of the deep ocean circulation, as you will see later on.

Question 1.6 The total amount of water on land at any one time approximates to $38\,000 \times 10^{15}$ kg (Figure 1.3). The input (or output) rate, i.e. precipitation (or evaporation and run-off), for land areas is 100×10^{15} kg per year. So:

$$\text{residence time} = \frac{38 \times 10^{18}}{100 \times 10^{15}} = 380 \text{ years}$$

This average will conceal considerable variations because run-off into streams and rivers is a great deal more rapid than the movement of glaciers or the infiltration and movement of groundwaters. Groundwater and water frozen in ice-caps will both have much greater residence times than the average of 380 years suggests.

Question 1.7 (a) Arctic ice is mainly sea-ice, formed when seawater freezes in the Arctic Ocean basin. Antarctic ice comes mainly from the vast ice-sheet covering the Antarctic continent. The area of ice round the continent is extended by the direct freezing of seawater in winter.

(b) According to Figure 1.2, the sample of water will (i) attain its maximum density at about −0.3 °C, and it will (ii) freeze at about −1 °C.

(c) Ice melting into seawater will add freshwater and will thus dilute the seawater, making it less saline. This will raise the freezing point of the seawater, and will therefore facilitate the formation of further sea-ice when temperatures fall once more.

CHAPTER 2

Question 2.1 The albedo of snow is greater than that of vegetation (Table 2.1), so a snowfield will reflect more solar radiation than a forest, i.e. it will *absorb* less. If it absorbs less short wave radiation, it will re-emit less long wave radiation back to the atmosphere. So, the atmosphere above a forest should be warmed more than that above snow-covered ground.

Question 2.2 The three reasons are (cf. Question 1.5):

1 Water is transparent, so the radiation penetrates some distance below the surface (but see Figure 2.5), and because water is a liquid, heat is carried to deeper levels by mixing (see Section 2.3).

2 Water has a much higher specific heat (Table 1.1) than rock or soil, and absorbs more heat for a given rise in temperature.

3 Water has very high latent heats of evaporation and fusion (Table 1.1). Large amounts of heat are required to achieve evaporation of water or melting of ice, without *any* rise in temperature.

Question 2.3 (a) From Figure 1.3, about $336 \times 10^{15}/365$ kg of water leave the oceans daily by evaporation, i.e. 920×10^{12} kg day^{-1}.

So, the heat lost per day is:

$$2.25 \times 10^6 \, \text{J kg}^{-1} \times 920 \times 10^{12} \, \text{kg} = 2.25 \times 920 \times 10^{18} \, \text{J} \approx 2.1 \times 10^{21} \, \text{J}$$

(b) 2.1×10^{21} J approaches one-quarter of the daily global insolation value, given as 9×10^{21} J, so the loss of heat by evaporation from the oceans is a very important component of the Earth's heat budget.

(c) When warm saturated air passes over a cold sea-surface, water vapour will condense on the surface, giving up its latent heat as it does so (water vapour condensing in the air above a cold sea-surface forms advection fog).

Question 2.4 (a) Profile I in Figure 2.6(b) corresponds to line A in Figure 2.6(a), and profile II matches with line B. In the equatorial profile (II), the thermocline is nearer the surface and has a steeper gradient (the isotherms are closer together) than the thermocline for the mid-latitude station (profile I).

(b) At high latitudes, there is no thermocline and a temperature profile would be a virtually straight vertical line (see Figure 2.7).

Question 2.5 The curve would probably lie mainly between 18 °C and 19 °C, and would probably be a more or less horizontal line.

Question 2.6 (a) Table 1.2 shows that freshwater reaches its maximum density at 4 °C, so a surface layer of water cooler than this would be less dense than underlying water at 4 °C. This is a gravitationally stable situation, and the less dense water cooler than 4 °C would remain at the surface. The presence of dissolved salts in seawater means that the density increases right down to freezing point (Figure 1.2). Seawater cooler than 4 °C will be more dense than underlying warmer layers. That is a gravitationally unstable situation and the colder water at the surface will sink – *unless* the underlying layers happen to be sufficiently more saline to be denser, an aspect we shall explore again in Chapter 4.

(b) The answer again lies in the maximum density of freshwater being at 4 °C. As soon as the temperature of the water falls below this, it becomes less dense and it cannot sink. Water at more than 4 °C is likewise less dense. The densest water in a freshwater lake will always be that which is at 4 °C, and for the lake to be gravitationally stable this water must be at the deepest part of the lake.

Question 2.7 In broad terms, as shown in Figure 2.9, we can recognize first the *upper warm layer*, in which seasonal change is most evident and in which seasonal and diurnal thermoclines develop during summer in mid-latitudes. Below this lies the *main* or *permanent thermocline*, where temperatures fall relatively quickly from more than 10 °C to about 5 °C. Finally, the bulk of the deep oceans (the *deep layer* from around 1 000 m to the bottom) is characterized by temperatures of less than about 5 °C. This three-layered structure is not seen in high latitudes where temperature profiles are nearly vertical (Figure 2.7(c)).

CHAPTER 3

Question 3.1 Table 3.1 shows ionic concentrations by weight and gives no indication of charge. Seawater is electrically neutral because the proportions of total positive and total negative ions in *molar* terms are equal. To work out the molar concentration of an ion, you simply divide the concentration in parts per thousand by the appropriate relative atomic or molecular mass. When we do this for Na^+ and Cl^-, we find that their ionic proportions (i.e. molar concentrations) are much closer than their actual concentrations:

$$Na^+ : \frac{10.6\,g\,l^{-1}}{23\,g\,mol^{-1}} = 0.46\,mol\,l^{-1} \quad Cl^- : \frac{19.0\,g\,l^{-1}}{35.5\,g\,mol^{-1}} = 0.53\,mol\,l^{-1}$$

You can verify the overall balance of the ions in Table 3.1 by doing the same for all the other positive and negative ions, and adding up the totals.

Question 3.2 Only four of the elements in Table 3.2 appear in Table 3.1: Ca, K, Na and Mg (though not in the same sequence of relative abundance).

Question 3.3 (a) From Table 3.1, the ratio is:

$$\frac{\text{concentration of } K^+}{\text{total salinity}} = \frac{0.380}{34.482} = 0.011$$

(b) (i) For a salinity of 36:

$$\frac{\text{concentration of } K^+}{36} = 0.011$$

concentration of K+ = 0.011 × 36 = 0.396 (‰ by weight)

(ii) and for a salinity of 33:

concentration of K+ = 0.011 × 33 = 0.363 (‰ by weight)

(c) From Table 3.1,

$$K^+ : Cl^- = \frac{0.380}{18.980} = 0.020.$$

This ratio should be the same in each of cases (i) and (ii) of part (b), and for any other value of salinity in the open oceans, because the relative proportions of major ions remain constant.

(d) Salinity will increase if water is withdrawn from the oceans by evaporation or by the formation of sea-ice; and it will decrease if water is added to the oceans by precipitation, run-off from rivers, or melting ice.

Question 3.4   If oxygen is taken from sulphate (SO_4^{2-}) ions, the sulphate will be reduced to sulphide (S^{2-}). The ratio of SO_4^{2-} to salinity will decrease, with respect to normal seawater.

Question 3.5   (a)   Profile I on Figure 3.3(b) corresponds to line A on Figure 3.3(a). Profile II corresponds to line B. The depth range of the halocline on each profile corresponds quite closely with that of the thermocline at the same location on Figure 2.6.

(b)   The halocline is nearer the surface for the equatorial profile (II), and the rate of decrease of salinity with depth across the halocline is much greater (i.e. the halocline is steeper) – you can tell that from Figure 3.3(a), because the isohalines near the surface are crowded together in the equatorial belt (compare isotherms along line B (profile II) of Figure 2.6).

(c)   No. At *high* latitudes, relatively low salinity values are found from the surface to the bottom (cf. Figure 3.3(a)), so the profiles would approximate to vertical lines.

Question 3.6   Surface salinities are controlled by the balance between evaporation and precipitation. Figure 3.4(b) shows how close this relationship is: where precipitation exceeds evaporation, the salinity is low; where evaporation is high, so is the salinity. The equatorial salinity minimum (see also Figure 3.3(a)) is due to an excess of rainfall over evaporation along the equatorial belt, despite high average insolation and temperatures.

Question 3.7   35. If R_{15} is equal to 1, equation 3.3 simplifies as follows:

$$S = 0.008\,0 - 0.169\,2 + 25.385\,1 + 14.094\,1 - 7.026\,1 + 2.708\,1 = 35.$$

Question 3.8   (a)   False. Tables 3.1 and 3.2 show that the proportions are quite different.

(b)   True. Chloride is a major constituent of seawater and its ratio to salinity is constant throughout most of the oceans. That is why chlorinity measurements were for a long time the principal means of determining salinity.

(c)   True. If Ca^{2+} ions are selectively removed from solution, then the $Ca:S$ ratio must fall.

(d)   Partly true. Haloclines can represent a decrease *or* an increase in salinity with depth. The main halocline typically represents a decrease (Figure 3.3).

(e)   False. Measurements of salinity and temperature are routinely quoted to ± 0.01 and $\pm 0.01\,°C$, but they can have a precision of ± 0.003 or better (Section 3.3.2).

Question 3.9   If dissolved salts are being continually added to the oceans and they nevertheless maintain a constant composition – both in the total amount and the relative proportions of dissolved constituents – then elements must be removed (precipitated) from seawater (by biological activity or chemical precipitation), *at the same rates* at which they are being added. If this were not the case, there would be much greater variations of salinity throughout the oceans and ratios of major elements both to one another and to total salinity would fluctuate a lot more than they do. We shall be exploring this in Chapter 6.

CHAPTER 4

Question 4.1 (a) The boundaries are sketched in on Figure A1. The three water masses are known as: 1, Antarctic Bottom Water (AABW); 2, North Atlantic Deep Water (NADW); 3, Antarctic Intermediate Water (AAIW). See also Section 4.3.3.

(b) (i) Within individual water masses, temperature, salinity and other properties are fairly uniform.

(ii) At boundaries between water masses, however, there can be quite sharp changes. The gradients of change in T and S across boundaries gradually become more blurred as water masses move further from their source regions (Figure A1) and sufficient time has elapsed for significant mixing to have occurred across the boundaries.

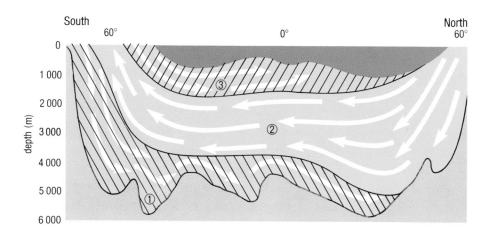

Figure A1 Answer to Question 4.1.

Question 4.2 (a) (i) Surface waters in polar regions are cooled both by direct contact with ice-sheets and by cold winds blowing off the ice. Sea-ice formed when seawater freezes leaves behind seawater of slightly higher salinity (Section 1.2.2). Both these processes combine to increase the density of surface waters (cf. Figure 1.2).

(ii) In tropical regions, the salinity (and therefore the density) of surface waters is increased mainly by evaporation (cf. Figure 3.4). Strong winds also cool surface waters and increase their density.

(b) Yes, of course: cold air at the poles extracts thermal energy from the surface of the ocean, either directly by blowing over it, or indirectly through contact with (and formation of) ice. The cold, dense water so formed sinks to the ocean floor. In lower latitudes, heating of the ocean surface ensures that the water remains generally less dense there. This heating contributes substantially to the total global hydrological cycle by causing evaporation. Some of the water evaporated in low latitudes is transported in the atmosphere towards polar regions, where the water falls as snow.

Question 4.3 (a) Substituting in the hydrostatic equation (equation 4.1):

pressure at 10 m depth, $P = 1.03 \times 10^3 \, \text{kg m}^{-3} \times 9.8 \, \text{m s}^{-2} \times 10 \, \text{m}$

$$\approx 10^5 \, (\text{kg m s}^{-2}) \, \text{m}^{-2}$$

$$\approx 10^5 \, \text{N m}^{-2}$$

According to Figure 4.3, a 10 m column of seawater exerts a pressure equivalent to one atmosphere, which can also be expressed as 1 000 millibars (mbar; 1 000 mbar = 1 bar).

(b) (i)   The deep ocean floors are around 4–5 km deep, which corresponds to pressures in the order of 4×10^7 to $5 \times 10^7 \, \text{N m}^{-2}$ or 400–500 atmospheres (400–500 bar).

(ii)   Some ocean trenches reach depths of about 10 km, so pressures there must be about $10^8 \, \text{N m}^{-2}$, or 1 000 atmospheres (1 000 bar).

Question 4.4   (a)   Air would be subjected to *increased* pressure in the adiabatic change from a height of 5 km down to sea-level. It would become more compressed, so its potential temperature will be *higher* than its *in situ* temperature.

(b)   Seawater would be subjected to *reduced* pressure in the adiabatic change from a depth of 5 km up to sea-level. It would become less compressed, so its potential temperature will be *lower* than its *in situ* temperature.

Question 4.5   (a)   For (i), σ_t is 27.6 kg m^{-3} (interpolating between the contours on Figure 4.4); for (ii), σ_t is 26.4 kg m^{-3}.

(b)   These two σ_t values 'translate' into density values of $1.027\,6 \times 10^3$ kg m^{-3} and $1.026\,4 \times 10^3$ kg m^{-3}, respectively.

Question 4.6   Figure 4.6(b) must correspond to the description because it shows coolest water at the bottom of the trench (i.e. the distribution of temperature unaffected by pressure), so its contours must represent potential temperature. Figure 4.6(a) shows temperature actually *increasing* below 4 000 m, so the contours must represent *in situ* temperature uncorrected for adiabatic compression.

Question 4.7   (a)   Most of the curve crosses the contours of equal density in such a way that it shows a consistent increase of σ_t with depth, and suggests that the water column is stable down to at least 2 000 m. Below that, the line is roughly parallel with the contours, suggesting a more or less constant density over the lowermost 3 000 m, which means that this is the least stable part of the water column (see the text following this question).

(b) σ_t can only give a rough indication of stability, as it is uncorrected for adiabatic heating (by definition), and so it gives spurious information about density (and hence stability), especially in deep water. A more reliable indication of stability would be obtained by plotting σ_θ against depth.

(c)   The curve would be similar in shape to that in Figure 4.7 but would be displaced progressively downwards with respect to the temperature axis, with increasing depth. The point at 150 m depth would be virtually the same for both curves, and the divergence between the curves would increase with depth, becoming greatest at 5 000 m, where comparison with Table 4.1 suggests that the potential temperature would be *c.* 0 °C.

Question 4.8   (a) Ca^{2+} and HCO_3^- participate in biological processes. Their concentrations are changed by processes other than mixing. They therefore behave non-conservatively.

(b)   Chlorinity must be a conservative property because it is used to calculate S, which is conservative. Within the main body of the oceans, chloride, iodide and bromide do not participate in processes that remove them from the seawater solution to any significant extent. Their concentrations are changed only by mixing.

Question 4.9   The density of seawater is increased both by direct cooling and by the removal of freshwater to form sea-ice, which increases the salinity of the seawater left behind. The top of the water column becomes gravitationally unstable, surface water sinks, and moves towards equatorial regions. The main sources of these cold water masses are in the Greenland and Norwegian Seas, the Labrador Sea and around Antarctica, particularly in the Weddell Sea.

Question 4.10   No. In both cases the temperature falls with increasing distance from the Earth's surface, where most of the insolation is absorbed (directly or indirectly). The resemblance ends there. Decrease of temperature with depth in the oceans is due partly to the limited penetration of sunlight (see also Chapter 5); and partly to the limited downward extent of vertical mixing. Temperatures at depth are also low because of the thermohaline circulation, which supplies cold water from polar regions to the deep oceans. Decrease of temperature with height in the troposphere is due mainly to the adiabatic expansion of air as it rises on being warmed at the Earth's surface.

Question 4.11   Density values of $1.025 \times 10^3\,\mathrm{kg\,m^{-3}}$ and $1.026 \times 10^3\,\mathrm{kg\,m^{-3}}$ are represented by σ_t values of $25.0\,\mathrm{kg\,m^{-3}}$ and $26.0\,\mathrm{kg\,m^{-3}}$. From Figure 4.4, if σ_t is to increase by the amount indicated, then either: (a) the temperature must decrease from $16\,^\circ\mathrm{C}$ to about $11\,^\circ\mathrm{C}$; or (b) the salinity must increase from 34.0 to about 35.3.

(c)   The definition of σ_t means that density is calculated as if the water sample were under atmospheric pressure. Water at a depth of 4 000 m will be under a pressure of some 400 atmospheres (Figure 4.3). So its *in situ* temperature will be higher than its potential temperature, and the density corresponding to its σ_t value ($27.6\,\mathrm{kg\,m^{-3}}$) is bound to be less than its potential density.

Question 4.12   (a)   The large spread of values must represent mainly upper water masses. Examination of any sections or profiles (e.g. Figure 2.6, Figure 3.3) shows that the largest ranges of T and S are in the upper 1 000 m of the water column.

(b)   The three values plot close together in the bottom of Figure 4.18 but there is a small spread which corresponds to that of the main fields shown on the diagram (i.e. Pacific on the left, Atlantic on the right, Indian in the middle). The *average* values for ocean waters will be determined largely by the great mass of water of relatively low temperature and salinity below the main halocline and the permanent thermocline.

Question 4.13   (a)   True. *In situ* temperature is subject to change not only by mixing, but also by adiabatic heating or cooling as a result of changes in pressure. Hence it does not fall within the strict definition of conservative properties. Potential temperature has been corrected for adiabatic changes, and is therefore a true conservative property.

(b)   False. Only the *in situ* temperature can be directly measured. Potential temperature is obtained by correcting the measured temperature for the effects of adiabatic compression.

(c)   True. Both processes lower the density of surface waters: warm water is less dense than cold water, and freshwater is less dense than seawater. Both processes strengthen the stratification and increase the stability of the water column.

(d)   False (see Section 2.3). Steep thermoclines and steep haloclines respectively represent rapid changes of temperature and salinity with depth and would thus characterize well-stratified water columns.

(e)   True. Density increases as temperature decreases, so for stability to be maintained here, the slightly warmer water at 250 m depth must have higher salinity than water at the surface.

Question 4.14   (a)   By extrapolation from Figure 4.4, (i) water at $-1\,°C$ and with salinity of 33 has σ_t of about $26.7\,kg\,m^{-3}$; while (ii) water at $-1\,°C$ and with salinity of 35 has σ_t of $28.2\,kg\,m^{-3}$. These correspond to densities of $1.026\,7 \times 10^3\,kg\,m^{-3}$ and $1.028\,2 \times 10^3\,kg\,m^{-3}$.

(b)   Water of salinity 33 would freeze at about $-1.8\,°C$, from Figure 1.2, but even as it approached this lower temperature its density would not increase above that of the underlying water of salinity 35. Sea-ice could therefore form.

CHAPTER 5

Question 5.1   (a)   (i) No. Even at the surface of the ocean, moonlight has an intensity about four orders of magnitude lower than that required for phytoplankton growth. (ii) No, normally the limiting depth in even the clearest coastal water in sunlight is about 50 m. (iii) Yes. Indeed, phytoplankton can grow only during daylight within this narrow depth zone, and if the water is not clear then the depth is less.

(b)   (i) No. Moonlight of the required minimum intensity for perception penetrates only to a depth of about 700 m. (ii) Yes, but this approaches the limit: below about 1 250 m, they must live in virtually total darkness.

Question 5.2   Dark coloration will reduce contrast when the fish is viewed from above or from the side, i.e. against a (normally) dark bottom and dark surrounding water. A silvery underside will similarly reduce the contrast when the fish is viewed from below, against surface waters illuminated by daylight, i.e. against a background of downwelling irradiance. In both cases, the fish is less obvious to both predator and prey.

Question 5.3   (a)   The coefficient of attenuation for directional light must always be greater than the coefficient of diffuse attenuation for non-directional light. An underwater surface can be illuminated by light from any direction, including that which has been appreciably scattered; whereas an object can only be perceived as a result of light that has travelled in a direct line from the object to the eye (or camera) – and most of the light leaving the object along the required path will be scattered away from that path.

(b)   Sunlight penetrates furthest through the clearest ocean water, so this must have the smallest coefficients of attenuation. Coastal water is generally more turbid and so has larger coefficients.

Question 5.4 (a) Substituting in equation 5.1:

$$10 = \frac{8}{C + K}$$

$$C + K = \frac{8}{10} = 0.8\,\mathrm{m}^{-1}$$

(b) Following Question 5.3(a) and related text, the attenuation coefficient, C, is greater than the extinction coefficient, K, and therefore contributes a larger proportion to the sum. The proportion will be greater in turbid water.

Question 5.5 (a) Z_S should be less where production is high, because large populations of phytoplankton in the water greatly increase the concentration of seston, and thus attenuate both directional and non-directional light much more than in regions where primary production is low.

(b) Z_S should be greater before, because a storm will greatly increase turbidity of the water by increasing the amount of suspended sediment both from the sea-bed and by run-off from land.

Question 5.6 Yes. In water that is relatively clear, blue–green light penetrates to greatest depth. Algae such as kelp look brown because their main pigment absorbs blue–green light and reflects light of other wavelengths. Of course, many brown seaweeeds grow *above* low-tide mark and in very shallow water. For this reason, they also contain chlorophyll (but its colour is masked by that of other pigments).

Question 5.7 The answer is *not* $3 \times 35 = 105\%$. If 35% is absorbed, then 65% is transmitted, and this becomes the 'incident light' for the next 1 m, and so on, because of the exponential relationship between illumination and depth (Figure 5.1). So, in 3 m, the transmission will be:

65% of 65% of 65% or $65 \times 0.65 \times 0.65 = 27.5\%$.

That is the proportion transmitted, so the proportion absorbed is $100 - 27.5 = 72.5\%$.

Question 5.8 The speed of sound in seawater is less than that in rock, but it is greater than that in air (see also Table 5.1). The reason for the apparent anomaly is that, in general, denser materials have higher axial moduli – they are 'stiffer' (i.e. less compressible); and the axial modulus appears in the numerator of equation 5.5.

Question 5.9 Substituting in equation 5.6:

$$c = 1\,410 + (4.21 \times 10) - (0.037 \times 10^2) + (1.14 \times 35) + (0.018 \times 100)$$
$$= 1490.1\,\mathrm{m\,s}^{-1}.$$

Question 5.10 If $Z_1 = Z_2$, then $Z_1 - Z_2$ is zero, and R must also be zero. Sound will cross the interface with little or no loss of acoustic energy.

Question 5.11 (a) The speed of sound in the sound channel changes significantly from one end of the section to the other. It is less than $1\,470\,\mathrm{m\,s}^{-1}$ at 50° S, reaches nearly $1\,500\,\mathrm{m\,s}^{-1}$ at 30° N and then declines again, to about $1\,480\,\mathrm{m\,s}^{-1}$.

(b) The depth of the sound channel axis is greatest at around 30° N, remains at around 1 km between about 10° N and 40° S, and becomes very shallow poleward of about 50° latitude. The main reason for the fluctuations is that the depth of the thermocline varies (cf. Figure 2.9); at higher latitudes the whole water column is well mixed, so the permanent thermocline and main halocline are absent (cf. Figures 2.6 and 3.3).

(c) From Section 5.2.2, increases in both T and S above the sound channel axis lead to an increase in the speed of sound. At depth, neither T nor S changes much, but pressure increases the axial modulus of the water more than it increases the density. So, from equation 5.5, c must increase.

Question 5.12 (a) Not very. A change in temperature of 1 °C leads to a change in sound speed of the order of 3 m s^{-1}, whereas a change in salinity of 1 changes the speed of sound by only 1.1 m s^{-1} (Section 5.2.2). A change in salinity of 0.1 thus affects the speed of sound hardly at all in comparison with a temperature change of a degree or two.

(b) Figure 5.9 shows that (i) attenuation of sound is less for low than for high frequencies, and range is more important than resolution in this kind of experiment, where sound is transmitted over long distances; (ii) at low frequencies the main cause of attenuation is inhomogeneities in the water column, so less well-mixed (less homogeneous) bodies of water could be distinguished by the greater attenuation of acoustic pulses transmitted through them.

Question 5.13 Because they have been absorbed in the topmost 10 m of the water column, cf. Figure 2.5.

Question 5.14 Some fish have an air-filled swim-bladder which occupies only about 5% of the total volume of the fish, yet may account for more than 50% of the returning echo, because of the very high reflectivity of air–water interfaces (Table 5.1). Indeed, air bubbles flowing along the hull of a heaving fishing boat can blanket sonar equipment and render the fish-finder device almost inoperable. (Cartilaginous fish such as sharks, rays, skates and dogfish do not have swim-bladders; nor do some bony fishes, such as mackerel and tuna. Swim-bladders of fishes that live below about 1 000 m are filled with fat, not gas, and swim-bladders of sea-bed-dwelling fish are used as auditory organs rather than for buoyancy, Section 5.2.3.)

Question 5.15 (a) To detect individual fish in a shoal or near the sea-bed requires good discrimination; so fishing sonars are high frequency, short range systems.

(b) Presumably, one would wish to detect a submarine at maximum possible range; however, the size of such a target would not require highly accurate discrimination. Antisubmarine sonars therefore operate at lower frequencies to maximize the range.

Question 5.16 (a) False. Open ocean water is clearer and more blue than nearshore water, which is often yellowish as well as turbid. Light penetration in coastal water is less and is biased away from the blue–green light required by most photosynthesizing organisms.

(b) False. The water is described as being of an 'intense blue', which means it is non-productive biologically (Section 5.1.4).

(c)   True. Water below the permanent thermocline is usually at less than 6 °C, which is the lower limit of temperature given for equation 5.6.

(d)   True, but does it matter? The point is that the acoustic impedance of seawater is so enormous compared with that of air (Table 5.1) – nearly four orders of magnitude greater – that for all practical purposes, the acoustic reflectivity of the air–sea interface is 100%.

Question 5.17   The broken line represents the mean winter profile of sound speed with depth. As temperature falls, the speed of sound decreases. (We are ignoring any effects of salinity changes, which generally have only a small influence on the speed of sound; see equation 5.6 and Question 5.12.) The depth of the sound channel axis is about 600–700 m, and changes at and below this depth are minimal because seasonal changes penetrate no more than a few hundred metres below the surface (cf. Figure 2.8).

CHAPTER 6

Question 6.1   Table 6.1 gives the concentrations of individual elements, whereas in Table 3.1 the concentrations are listed in terms of predominant ionic *species*, i.e. the form in which the dissolved constituents occur (see Section 6.3.1). For the simple ions, e.g. Na^+ and Cl^-, ionic and elemental concentrations by weight are the same, but for most anions they are clearly different. Sulphur occurs predominantly as the sulphate anion (SO_4^{2-}). Carbon is present as dissolved carbon dioxide gas (CO_2), as carbonic acid (H_2CO_3) and its dissociation products (HCO_3^- and CO_3^{2-}); together with a small but rather variable amount (about 1 mg l^{-1}) in the form of dissolved organic molecules. Boron occurs as the hydroxide ($B(OH)_3$) and related ionic forms. Presenting analyses in molar terms avoids such disparities because, for example, a mole of bicarbonate ion (HCO_3^-) contains a mole of carbon (C); a mole of sulphate ion (SO_4^{2-}) contains a mole of sulphur (S); a mole of boron hydroxide ($B(OH)_3$) contains a mole of boron (B); and so on; so the concentrations become the same.

Question 6.2   For this calculation:

$$g = 9.8 \text{ m s}^{-2}$$
$$d = 2 \times 10^{-6} \text{ m}$$
$$\rho_1 = 1.5 \times 10^3 \text{ kg m}^{-3}$$
$$\rho_2 = 1.0 \times 10^3 \text{ kg m}^{-3} \text{ (close enough to seawater for our purposes)}$$
$$\mu = 10^{-3} \text{ N s m}^{-2} \text{ (close enough to seawater for our purposes)}$$

Then, substituting in equation 6.1,

$$v = \frac{1}{18} \times 9.8 \times \frac{(1.5 - 1.0) \times 10^3}{10^{-3}} \times (2 \times 10^{-6})^2$$

$$= \frac{9.8}{18} \times 0.5 \times 10^6 \times 4 \times 10^{-12}$$

$$= \frac{9.8 \times 2}{18} \times 10^{-6} = 1.09 \times 10^{-6} \text{ m s}^{-1}$$

This particle would therefore take about $1/(1.09 \times 10^{-6}) = 0.92 \times 10^6$ seconds (about 255 hours or more than 10 days) to sink through 1 m.

Question 6.3   Well-stratified waters are gravitationally stable. Nutrients which sink from the surface cannot be replaced except by slow diffusion processes. If the upper water column is well mixed, on the other hand, nutrients sinking from the surface have a better chance of being carried back up again.

Question 6.4   From Figure 6.4, the ratio of $N_2 : O_2$ in air is close to $8 : 2$ or $4 : 1$, whereas in seawater in equilibrium with the atmosphere it drops to about $9 : 5.3$ or $1.7 : 1$. As $4 \div 1.7 \approx 2$, the solubility of oxygen is more than double that of nitrogen.

Question 6.5   (a)   Chiefly because the source regions are very cold, and Figure 6.5 shows the solubility of oxygen to increase with falling temperature. The oceans are also rougher at high latitudes and wave-breaking encourages solution of gases.

(b)   Oxygen concentrations are lowest in profile I: water is almost anoxic in the oxygen minimum layer. This could be due partly to the respiratory needs of large populations of animals supported by high levels of primary production in surface waters, but perhaps mainly to bacterial activity associated with the decomposition of organic detritus sinking out of the photic zone.

Question 6.6   The sea $\rightarrow$ air flux of N_2O from Table 6.2 is $1.2 \times 10^{14}\,\text{g yr}^{-1}$. That is:

$$1.2 \times 10^{14} \times \frac{28}{28 + 16}\ \text{gN yr}^{-1}$$

$$= 1.2 \times 10^{14} \times 0.64$$

$$= 7.7 \times 10^{13}\ \text{gN yr}^{-1}$$

The balance to be made up is:

$$(8.0 - 0.9) \times 10^{13}\ \text{gN yr}^{-1} = 7.1 \times 10^{13}\ \text{gN yr}^{-1}$$

The sea $\rightarrow$ air flux of N_2O gas more than makes up this balance. In fact, the agreement is extremely good, given that global estimates of fluxes of this nature are subject to considerable uncertainties.

Question 6.7   (a)   Oxygen is non-conservative, because its concentration is changed by biological or chemical reactions, i.e. not by mixing only. The concentration of dissolved oxygen in subsurface water masses decreases with time and with distance from the source regions, because it is used up in biological respiration and microbial decomposition. Below the photic zone there is no mechanism of replenishment other than downward mixing from the surface, and upward mixing from the deep ocean (cf. Figure 6.7). Both processes are too slow to keep pace with the rate of consumption.

(b)   If you compare Figure 6.7(a) with Figure A1, you should be able to see the long 'tongue' of North Atlantic Deep Water, extending southwards at mid-depths. The overlying Antarctic Intermediate Water is also visible. Less obvious is the gradual decrease in dissolved oxygen content from surface to bottom in Antarctic regions that corresponds to sinking Antarctic Bottom Water.

(c)   Figure 6.7(b) shows that in the North Pacific there is no sinking of surface waters rich in oxygen; instead there is a strong oxygen-minimum layer at *c*. 500–1 000 m depth throughout most of the North Pacific (where much of the water is sub-oxic). The dissolved oxygen contours thus give no indication of a source region of deep water in the North Pacific, which (as noted in Section 4.1) is effectively isolated from the Arctic Ocean by the shallow barrier of the Aleutian Islands chain.

Question 6.8 (a) (i) Rainwater is about 5 000 times more dilute than seawater: $(34.4 \, \mathrm{g} \, \mathrm{l}^{-1})/(7.1 \times 10^{-3} \, \mathrm{g} \, \mathrm{l}^{-1})$; (ii) river water is about 300 times more dilute than seawater: $(34.4 \, \mathrm{g} \, \mathrm{l}^{-1})/(118.1 \times 10^{-3} \, \mathrm{g} \, \mathrm{l}^{-1})$.

(b) The similarity in composition between rainwater and seawater is particularly striking. Most of the dissolved salts in rainwater have a marine origin, as a consequence of the injection of seawater into the atmosphere as aerosols produced by bubble-bursting at the sea-surface (Section 2.2.1). River water is very different in composition, with relatively much more Ca^{2+} and HCO_3^- and SiO_2, and much less Na^+ and Cl^-.

Question 6.9 Clearly (i) and (iv) must apply to seawater, and (ii) and (iii) to river water corrected for cyclic salts.

Question 6.10 The amount of average crustal rock that must be weathered to provide 11 g of sodium in solution is:

$$\frac{11}{1.8} \times 100 \approx 600 \, \mathrm{g}$$

Question 6.11 (a) Sodium, potassium, magnesium and chloride have residence times that are changed significantly by making the correction for cyclic salts.

(b) If $2.5 \times 10^8 \, \mathrm{t} \, \mathrm{yr}^{-1}$ are added to the $4.88 \times 10^8 \, \mathrm{t} \, \mathrm{yr}^{-1}$ in Table 6.4, the uncorrected residence time for calcium becomes significantly shorter:

$$\frac{\text{mass of Ca in oceans}}{\text{rate of supply of Ca to oceans}} = \text{residence time}$$

$$\frac{6 \times 10^{14}}{7.38 \times 10^8} \approx 800\,000 \text{ years (i.e. } 0.8 \times 10^6 \, \mathrm{yr})$$

The same would apply to the residence time corrected for cyclic salts.

Question 6.12 (a) The stirring (or mixing) time is the period for an average water molecule to travel from the surface to the deep ocean and back again (about 500 years). The residence time is the period (about 4 000 years) spent by an average water molecule actually within the oceans, before being returned to the atmosphere in the hydrological cycle (Figure 1.3).

(b) Not many elements in Figure 6.11 have residence times of 500 years or less. However, iron and aluminium appear to be in the seawater solution for too short a time to permit complete mixing throughout the whole ocean. Other elements with residence times of less than about 10^3 years may also not be completely mixed.

Question 6.13 Mg^{2+} will have the largest hydration sphere relative to its size, as it has the greatest charge density of the three (smallest radius, biggest charge). Chloride will have the smallest hydration sphere relative to its size, as it has the lowest charge density of the three (largest radius, single charge).

Question 6.14 (a) Cl⁻ is missing from Table 6.5. It does not form an ion pair with any major cation. In fact, equilibrium constants for interactions between major cations and Cl⁻ (and, incidentally, between K^+ and HCO_3^- and CO_3^{2-}, see Table 6.5) show that association is not significant in these cases. In other words, these ions behave effectively as free ions with respect to one another, as in Figure 6.13(a).

(b) Chloride is the most abundant ion in seawater, and as it does not form ion pairs, this 'burden' falls on the remaining anions, which are present in much lower concentrations. A greater proportion of these anions are therefore involved in ion-pair formation.

(c) $MgSO_4$ appears to be the most abundant. You encountered it in Chapter 5 (Figure 5.9), in the context of the attenuation of acoustic energy in the oceans.

Question 6.15 (a) Substituting in equation 6.11:

For surface sample:

$$2.35 - 2.0 = [CO_3^{2-}] = 0.35 \text{ mol m}^{-3}.$$

For deep sample:

$$2.55 - 2.4 = [CO_3^{2-}] = 0.15 \text{ mol m}^{-3}.$$

Substituting in equation 6.10:

For surface sample:

$$2.0 = [HCO_3^-] + 0.35, \text{ so } [HCO_3^-] = 1.65 \text{ mol m}^{-3}.$$

For deep sample:

$$2.4 = [HCO_3^-] + 0.15, \text{ so } [HCO_3^-] = 2.25 \text{ mol m}^{-3}.$$

Substituting in equation 6.15:

For surface sample:

$$[H^+] = 1.0 \times 10^{-9} \times \frac{1.65}{0.35} = 4.7 \times 10^{-9} \text{ mol l}^{-1}$$

For deep sample:

$$[H^+] = 1.0 \times 10^{-9} \times \frac{2.25}{0.15} = 1.5 \times 10^{-8} \text{ mol l}^{-1}$$

From equation 6.12 and the Appendix, $pH = -\log_{10}[H^+]$, so:

For surface sample:

$$\log 4.7 \times 10^{-9} = -9 + 0.7 = -8.3, \text{ and pH} = 8.3.$$

For deep sample:

$$\log 1.5 \times 10^{-8} = -8 + 0.2 = -7.8 \text{ and pH} = 7.8.$$

This example illustrates the general case: deep water in the ocean is generally more acid (lower pH) than surface water.

(b) If A is constant but $[\Sigma CO_2]$ increases, then $[CO_3^{2-}]$ must decrease, from equation 6.11. If $[\Sigma CO_2]$ increases, and $[CO_3^{2-}]$ decreases, then from equation 6.10, $[HCO_3^-]$ must increase. So the ratio of $[HCO_3^-] : [CO_3^{2-}]$ must also increase, and from equation 6.15, $[H^+]$ must increase, pH must fall, and the deep water must be more acid.

Question 6.16 (a)   Seawater is normally an oxidizing medium. Figures 6.5 to 6.7 show that there is free oxygen gas dissolved in the water. More generally, we know that the oceans are full of living organisms that need oxygen for respiration.

(b)   As normal seawater is an oxidizing medium, dissolved iron must be in its less soluble form, Fe^{3+}(aq).

(c)   The concentration should be higher in normal seawater, because uranium is more soluble in its oxidized form (U^{6+}).

Question 6.17 (a)   The straight line on Figure 6.17(a) has been extrapolated through the origin. The inference is that when nitrate is exhausted by biological activity, so is copper. Therefore, in these Antarctic waters at least, it would appear that copper could be biolimiting, as we know nitrate to be (Section 6.1.2, Figure 6.1(b)).

(b)   The nickel profile in Figure 6.17(b) has similarities to those for both phosphate and silica. We could therefore infer that biological processes influence nickel concentrations.

Question 6.18 (a)   Nitrate (NO_3^-) is almost totally depleted in surface water, and is obviously biolimiting.

(b)   Barium (Ba^{2+}) shows partial depletion in surface water and is bio-intermediate.

(c)   Sodium (Na^+) maintains a constant ratio of concentration to total salinity throughout the depth range and is therefore a bio-unlimited constituent.

Question 6.19 (a)   Residence time is given by: (mass in oceans)/(rate of input or removal). The greater the rate, the shorter the residence time. In this case, the rate of removal is the greater, and gives the shorter residence time.

(b)   Hydrothermal solutions are an important additional supply of dissolved constituents, including manganese, to seawater (Section 3).

Question 6.20   Anoxic conditions are reducing conditions, so manganese should exist in its Mn(II) form, i.e. as Mn^{2+} ions in solution.

Question 6.21 (a)   False. Only a very small proportion of the nitrogen in seawater (about 0.5 p.p.m., out of a total of 11.5 p.p.m.) is in the form of nitrate (Section 6.1.2).

(b)   False for most dissolved gases, because they are involved in chemical and/or biological processes that change their concentrations. Some inert gases could behave conservatively.

(c)   True. The (oxygen) compensation depth is where there is a balance between consumption of oxygen by respiration of plants and production of oxygen by photosynthesis. The oxygen minimum layer is where maximum net abstraction of oxygen has occurred, at around 500–1 000 m depth (Section 6.1.3).

(d)   False. There is plenty of dissolved oxygen in the photic zone (e.g. Figure 6.6). The limiting factor is light intensity, cf. Figure 5.1.

(e)   True. The greater the concentration in river water, the greater the annual supply from rivers relative to the total mass in the oceans, and the shorter the residence time.

(f)   False. Ca^{2+} has the greater charge density and will therefore have the relatively larger hydration sphere.

(g)   True. Salts will only be precipitated when they reach saturation. Figure 3.1 shows that large amounts of water must be evaporated to make the seawater solution more concentrated, so that saturation is reached (note that $CaCO_3$ is the exception to this – Section 6.3.2).

(h)   True. Hydrogen ions would be added, and from equations 6.7 and 6.8 the result must be to expel CO_2 gas from solution: reaction 6.2 moves to the left.

Question 6.22   It implies that the $Ca^{2+} : S$ ratio must be greater in the deep Pacific than in the deep Atlantic, because alkalinity can *only* be increased by dissolution of $CaCO_3$ (Figure 6.15).

CHAPTER 7

Question 7.1   The CO_2 in calcium carbonate accumulated in deep-sea sediments and precipitated in oceanic crust during hydrothermal circulation will eventually be recycled by subduction, re-melting and volcanism. Carbon in organic matter preserved in sediments is in chemically reduced form. Heating during subduction converts the organic matter to hydrocarbons, including methane gas.

Question 7.2   On an ice-free Earth, large amounts of cold dense water would not form in polar regions. Warm water contains less dissolved gas than cold water, so oxygen concentrations should be lower when average temperatures are high. Deep water would therefore be relatively oxygen-poor. However, there might well be a density-driven deep circulation on a warmer Earth, perhaps even quite vigorous, with high-salinity dense water sinking in regions of strong evaporation.

Question 7.3   If $[\Sigma CO_2]$ was greater, A would have been greater, too (equation 6.9). pH depends on the $[HCO_3^-]/[CO_3^{2-}]$ ratio (equation 6.15), which in turn depends on the size of the term $(A - [\Sigma CO_2])$ in equation 6.11. We cannot tell if this difference was small (low $[CO_3^{2-}]$, low pH, more acid) or large (high $[CO_3^{2-}]$, high pH, less acid). The early ocean may even have been relatively alkaline (high pH,), resembling East African 'soda lakes' that have high concentrations of bicarbonate and carbonate ion.

Question 7.4   Figure 7.3 shows that maximum insolation occurs: (a) in northern and southern mid-latitudes during the respective summer solstices (Figure 2.2), when days are long and daily values reach more than $25 \times 10^6\,J\,m^{-2}$; (b) along the Equator, where the average insolation is greater than $20 \times 10^6\,J\,m^{-2}\,day^{-1}$ (the average at mid-latitudes over the year is little more than $15 \times 10^6\,J\,m^{-2}\,day^{-1}$).

Question 7.5   The concentration in the mid-1990s was approaching 360 p.p.m. by volume (Figure 7.5), compared with less than 300 p.p.m. 130 000 years ago (Figure 7.2).

Question 7.6   (a)   Depending on which definition you prefer, alkalinity is represented by bicarbonate and carbonate ions (equation 6.9) or by the excess of strong base cations over strong acid anions (equation 6.16).

Equation 7.1 shows that weathering consumes hydrogen ions (acid) and releases bicarbonate (and carbonate) ions and cations (alkalinity) into solution. Reverse weathering removes bicarbonate (and carbonate) ions and cations (consuming alkalinity) and releases hydrogen ions (acid) back into solution.

(b)   Dissolution of $CaCO_3$ may be represented by the 'weathering' component of reaction 7.1, i.e. it releases alkalinity – which is why alkalinity is greater in deep than in surface waters.

(c)   Not by itself, cf. Figure 6.15. Alkalinity will be changed only if the overall balance of strong acid cations to strong base anions is changed (equation 6.16).

Question 7.7   In a steady-state ocean, if the rate of input of a dissolved constituent increases, then (a) the rate of removal must also increase, and (b) the residence time must decrease, because the concentration and hence the total mass in the ocean should not change.

Question 7.8   (a)   As cool water from below the thermocline rises to the surface by upwelling, pressure decreases and it becomes warmer. It is richer in total dissolved carbon than surface water (Figure 6.14(a)), and the effect of reduced pressure and warming is to release CO_2 gas from solution (as when you open a warmed bottle of carbonated mineral water, though less spectacular). Thus, low-latitude regions of upwelling can actually vent CO_2 to the atmosphere. In contrast, where deep water masses are forming, atmospheric CO_2 dissolved in cold water sinking from the surface is transported downwards to greater depths, where increased pressure ensures that it remains in solution (cf. reaction 6.2).

(b)   Upwelling brings nutrients to the surface and increases primary production (Section 6.1.2), which fixes more CO_2 in organic tissue. The more CO_2 fixed, the less escapes to the atmosphere.

Question 7.9   (a)   Cutting off the formation of NADW would cut off a major source of supply of dissolved CO_2 (and oxygen) to the deep ocean (cf. Figure A1), because dissolved atmospheric gases would no longer be carried downwards by cold dense water sinking from the surface (cf. Question 7.8(a)).

(b)   A 'lid' of low salinity melt water would greatly increase stratification at the top of the water column. This would tend to inhibit mixing and upwelling of nutrients from deeper water, so that phytoplankton production – and hence biological removal of CO_2 – would decline.

ACKNOWLEDGEMENTS

The Course Team wishes to thank the following: Dr Martin Angel and Dr Derek Pilgrim, the external assessors; Mr Mike Hosken and Mrs Mary Llewellyn for advice and comment on the whole Volume; Dr Derek Pilgrim for major contributions to Chapter 5, and Dr Ralph Rayner for help with Section 5.2.1. Dr Malcolm Howe and Dr Chris Vincent also provided helpful advice on content and level. For this second edition, the Course Team wishes to thank also: Dr Martin Angel, Mrs Sue Greig and Dr Simon Wakefield, the external assessors, and the many students, tutors and other readers who contributed comments and suggestions.

The structure and content of this Volume and of the Series as a whole owes much to our experience of producing and presenting the first Open University Course in Oceanography (S334), from 1976 to 1987. We are grateful to those people who prepared and maintained that Course, and to the students and tutors who provided valuable feedback and advice.

Grateful acknowledgement is also made to the following for material used in this Volume:

Figures 1.4 and 1.7 British Antarctic Survey; *Figures 1.5, 1.6 and 2.3* NASA; *Figure 2.4(a)* R. A. Horne (1969) *Marine Chemistry*, Wiley; *Figures 2.5 and 2.11* A. N. Strahler (1963) *Earth Sciences*, Harper and Row; *Figures 2.7, 4.5(a), 4.6 and Table 4.1* G. L. Pickard and W. J. Emery, *Descriptive Physical Oceanography – An Introduction*, 4th edn, Pergamon Press; *Figure 2.6(a), 2.8, 3.3(a), 4.1, 5.7, 6.7 and Table 1.1* H. U. Sverdrup *et al.* (1942) *The Oceans*, Prentice-Hall; *Figure 2.10(a)* US Department of Energy; *Figure 2.10(b)* G. Haber (1977) in *New Scientist*, **73,** New Science Publications; *Figure 3.2(a)* Centro de Caridade, Nossa Senhora do Perpetuo Socorro, Oporto; *Figure 3.4(a)* R. V. Tait (1968) *Elements of Marine Ecology*, Butterworths; *Figure 4.2* Estate of Mrs J. C. Robinson; *Figures 4.9–4.11* M. C. Gregg (1973) in *Scientific American*, **228**, W. H. Freeman; *Figure 4.14(a)* M. Hosken; *Figure 4.15* R. H. Stewart (1985) *Methods of Satellite Oceanography*, Scripps Institution of Oceanography/EROS Data Center; *Figure 4.16* R. C. Spindel (1982) in *Oceanus*, **25**, Woods Hole Oceanographic Institution; *Figure 4.17* NASA; *Figure 4.18* K. K. Turekian (1976) *Oceans*, 2nd edn, Prentice-Hall; *Figure 5.1* R. S. Dietz (1969) in *Readings in the Earth Sciences*, Volume 2, W. H. Freeman; *Figure 5.4 (part)* N. B. Marshall (1954) *Aspects of Deep-Sea Biology*, Hutchinson; *Figure 5.4 (photos)* P. M. David (upper), P. Herring (lower), IOS Deacon Laboratory (NERC); *Figure 5.6* Photo: Steve Johnson, Media Servies, University of Plymouth; *Figure 5.9* US Government Printing Office; *Figures 5.10 and 5.11* D. G. Tucker and B. K. Gazey (1966) *Applied Underwater Acoustics*, Pergamon; *Figure 5.13* Courtesy Defence Research Agency, Portland; *Figures 5.14 and 5.15* J. Northrop and J. G. Colborn (1977) in *Journal of Geophysical Research*, **79**, American Geophysical Union; *Figure 6.1(d) and (e)* Norman T. Nicholl; *Figures 6.2(b) and 6.3(a)* R. W. Jordan and M. Smithers; *Figure 6.3(b)* D. G. Jenkins; *Figure 6.3(c)* J. D. Milliman (1974) *Marine Carbonates*, Springer-Verlag; *Figure 6.5* W. S. Broecker and T. S. Peng (1982) *Tracers in the Sea*, Lamont-Doherty Observatory; *Figure 6.6* H. Friedrich (1969) *Marine Biology*, Sidgwick and Jackson; *Figures 6.11 and 7.1* M. Whitfield (1982) in *New Scientist*, 1 April; *Figure 6.14(a)* W. S. Broecker (1974) *Chemical Oceanography*, Harcourt, Brace, Jovanovich, Inc; *Figure 6.16* K. J. Orians and L. A. Bruland (1986) in *Nature*, **333**, Macmillan Journals; *Figure 6.17(b)* F. Sclater *et al.* (1976) in *Earth and Planetary Science Letters*, **31**, Elsevier; *Figure 7.2* J. Jouzel *et al.* (1993) in *Nature* **364**, Macmillan Journals; *Figure 7.6* W. Dansgaard *et al.* (1993) in *Nature* **364**, Macmillan Journals.

INDEX

*Note: page numbers in italics refer to illustrations; those in **bold** refer to terms being introduced or defined.*